BRISTOL
BEAUFIGHTER

BRISTOL BEAUFIGHTER

VICTOR BINGHAM

Airlife
England

Line drawings by Lyndon Jones MCSD

Copyright © 1994 by Victor Bingham

First published in the UK in 1994
by Airlife Publishing Ltd

British Library Cataloguing in Publication Data
A catalogue record for this book
is available from the British Library

ISBN 1 85310 122 2

Printed by Livesey Limited, Shrewsbury

Airlife Publishing Ltd.

101 Longden Road, Shrewsbury SY3 9EB, England

Contents

Acknowledgements

No factual book can be compiled without the help of many people who were, or are, involved in the presentation of information, and this book is no different. In my research on the Beaufighter I have received a great deal of help from staff at official establishments, from those who were involved with the aircraft, and from friends and acquaintances.

Accordingly, I would like to thank in particular the following, without whose help this book would not have been completed; Richard Searle and staff of Main Library/Archives RAE Farnborough; Brian Kervell, curator of the RAE Farnborough Museum; Paddy P.W. Porter for information and photographs from his collection; Fred Onions for research in Australia; Frank Smith for information and photographs on the RAAF; Alan G. Shepherd for information on radar installations; R. Mack of the RAF Museum Photographic Department; Frank Lovett for information on the Middle East; Group Captain John Wray for night-fighter information; Squadron Leader K. Lusty for the radar operator's point of view; Wing Commander J.G. Lingard; Flight Sergeant D.W. Gardner RAAF; R. Coombes; G. Muirhead; Keith Meggs; and R. Sherburn. Also a special thanks to Alastair Simpson of Airlife Publishing for guidance on the subject selection.

My thanks also to my wife for her help and patience, as well as being unofficial, unpaid secretary; also to my illustrator, Lyndon Jones, who manages to create drawings from my brief information.

Victor F. Bingham

Preface

The story of the Beaufighter has been told before, either from the point of view of the aircrew, or as part of a wider canvas. To the general public it is not as well known in detail as are the Mosquito, Spitfire and Lancaster; yet the Beaufighter was not only a very successful aircraft, but also formed the equipment of over forty squadrons in various RAF Commands, four squadrons in the USAAF, as well as a number of squadrons in the RAAF, RNZAF, SAAF and RCAF during the war years. So I make no apology for writing on an old subject, for the text will cover the design, development and testing periods, its construction and production, as well as a review of the Beaufighter's squadron service — and for me the Beaufighter is one of the most outstanding aircraft of World War II. As a design the Beaufighter was hardly the recipe for success, for it was an amalgam of parts of a not very successful torpedo bomber, allied to a new half fuselage and centre section, with a few more parts added, and powered by two Bristol Hercules engines.

After a few hiccups the Beaufighter was to go from success to success, in many roles and scenes, yet was to be overshadowed by the Mosquito when that aircraft entered service. It had its vices and virtues, as I have attempted to illuminate, yet did not suffer a spate of structural failures as did some aircraft. It needed to be mastered, to be flown in the prescribed manner — it was a strong, safe aircraft that did its job superbly. Apart from its various duties in Northern Europe, the Beaufighter became the scourge of shipping in the Mediterranean and Aegean Seas, a night-fighter and strike-fighter in North Africa, and an Army support and strike aircraft in Burma and along the eastern seaboard of the Indian Ocean. The Australians built it and used it in their strafing and bombing attacks on the Japanese enemy in the Pacific. To the Japanese the Beaufighter was known as 'Whispering Death' on account of its relatively quiet approach; and whereas the Mosquito in Europe was nicknamed the 'Wooden Wonder', I think it would be fair to state that in the Far East it never achieved the reputation of the Beaufighter — for the old Beaufighter just kept on going — at least two Beaufighter squadrons in SEAC were against changing them for Mosquitoes, and one squadron re-equipped with Beaufighters again after a period on Mosquitoes.

Climbing into the Beaufighter's heavily armoured cockpit, flanked on either side by a big, beautiful air-cooled Bristol Hercules engine, one had the feeling of the strength of the aircraft. When those two Hercules engines started up she became alive, she vibrated with power; and sitting in the cockpit with four 20 mm cannon stowed in her belly she appeared tough — she was tough. The 'Mighty Beau' was no sylph-like creature, no sleek flatterer, she looked what she was — a pugnacious beauty of a war machine, a tough lady — and I remember her that way. She filled a niche when nothing else was available; she operated in an enemy environment where the enemy held superiority, and her crews were forced to learn lessons and gain experience the hard way to the benefit of newer types of aircraft and crews.

Whilst others are satisfied to write about the aircrew under the title of the aircraft type, ignoring the technicalities and background, I feel that this fails to do justice to either the aircraft or title. So in this book I have attempted to indicate the trials and tribulations, the construction and achievements that lay behind the heroics, that were achieved on the battlefields where the Beaufighter roamed; whether it was over the grey-green seas of the north, the countries of Europe or the jungles of Burma. Behind every aircrew were numerous groundcrew, hundreds of workmen and designers at the factories, each playing their part in the web of placing the Beaufighter on the target; so to those who designed, manufactured, serviced and flew the 'Mighty Beau' I dedicate this book; may she and they be long remembered. Whether known as the 'Beau', 'Whispering Death' or 'Mighty Beau', she was the most versatile weapon-carrier in the RAF during World War II.

Vickers Venom, powered by the Bristol Aquila engine of 625 bhp. The smallest competitor, but it had a maximum speed of 312 mph.

Chapter 1
Specification and Design

Fighter design trends

The pace of aeronautical design and development had, by the mid-1930s, become so great, that the fighter design trend had positively finalized on the monoplane layout. Although up to 1930 there had been in the Air Ministry a deep-rooted prejudice of many years standing against the monoplane, now the time had arrived when an impartial examination and evaluation of the monoplane fighter concept would be carried out. This indicated that the thick monoplane wing offered an increased stowage for fuel, guns and ammunition, with a positive decrease in drag. During the same period a review of armament indicated a general increase in the number of guns that a fighter carried, consideration being towards a heavier calibre armament, mainly in the 20 and 23 mm range.

This period in Great Britain commenced in 1931, when the Air Staff issued Specification F7/30 to the aircraft industry for a single-seat day and night fighter. Compared to the contemporary fighters of that period, amongst other requirements were the following:

  Higher performance
  Exceptional manoeuvrability
  Long endurance
  Improved all-round vision
  Steep climb out on take-off
  Low landing speed
  Four machine-guns

The engine to be installed under the specification, indicated a preference by the Air Staff for the Rolls–Royce Goshawk evaporative cooled engine, which was another development of the Kestrel engine. The aircraft that were tendered to this specification were:

  Blackburn F7/30
  Bristol 123 and 133
  Hawker PV3

Supermarine 224
Westland PV4
Vickers 151 Jockey

Although testing and delays continued on up to 1935, the issue of a production contract and order eventually went to Gloster's private venture biplane, the Gladiator, under Specification F14/35.

The next specification to be drafted was F5/34, for Specification F35/35 for a high-speed single-seat fighter failed to produce anything but paper projects. F5/34 was the first specification which would bring in the monoplane fighter, carrying six or eight machine-guns, retractable undercarriage, and an enclosed cockpit and oxygen supply for the pilot. Competing for F5/34 were the Bristol type 146, Gloster G38, Martin-Baker MB2 and the Vickers Venom. Although all four aircraft were manufactured and flown, none reached production status, yet all were interesting designs in various ways. The Bristol 146 was powered by a Bristol Mercury of 840 bhp, carried eight machine-guns and had a maximum speed of 287 mph. The Gloster G38, also powered by a 840 bhp Mercury engine, recorded a maximum speed of 315 mph at 16,000 ft. The MB2, powered by a Napier Dagger engine of 798 bhp, carried eight machine-guns, had a fixed undercarriage yet achieved a maximum speed of 350 mph. The Vickers Venom was not only the smallest and squarest of the contenders, it was also the smallest powered entry, being pulled along by a Bristol Aquila of 625 bhp, yet had a top speed of 312 mph. Whilst the cockpit and gun installation of the Martin-Baker MB2 won unstinted praise at the A&AEE, criticism was voiced about the aircraft's stability.

After the impasse of the F7/30 specification, S. Camm of Hawkers proposed a low-wing monoplane version of the Hawker Fury, but

The Martin-Baker MB2, powered by a Napier Dagger III engine of 798 bhp, carried eight machine-guns and had a maximum speed of 350 mph.

powered with the Goshawk engine. With the introduction of the Rolls-Royce PV12 engine (later to be named the Merlin), his proposal was shelved and Camm began the detail design of his Interceptor Monoplane, which partly conformed to the F5/34 specification, but was a private venture. During this same period R. Mitchell of Supermarine was redesigning his type 224, and like Camm still using the Goshawk engine; then came the news of the PV12 engine, and so further redesign took place to incorporate this engine, the airframe then being designated the Type 300. These two private venture fighter designs would become the Hurricane and Spitfire respectively, with new specifications, F36/34 and F37/34 respectively. Before finalization of the two aircraft, the Air Staff carried out a further review of their armament installation, and with the licence agreement between Colt and BSA over the manufacture of the Browning 0.303 in. machine-gun, the specifications were amended to cover the installation of eight machine-guns.

Aircraft armament trends

During the period under review, British aircraft armament had up to 1930 not developed from the 1918 period, the Vickers 0.303 in. and Lewis 0.303 in. machine-guns still being the main armament. There had been a short flirtation with the Coventry Ordnance Works 1½ pdr (COW) gun as a fighter weapon, but the general installation was two rifle-calibre machine-guns. At the start of the 1930s a decision was made to test various machine-guns, including foreign weapons; this was because the machine-gun at that time in use had a low rate of fire and suffered from stoppages. With the proposed remote control installations in the new fighters, a machine-gun was required with a lower stoppage rate, and if the weapon selected was foreign, then a licence to produce it would be required. The result of the competition was the selection of the Colt (Browning) 0.300 in machine-gun, which would be bored out and redesigned to take rimmed 0.303 in ammunition, production initially to be in the hands of BSA.

In mainland Europe the aircraft armament trend was towards the 20–25 mm cannon, with Hispano-Suiza, Hotchkiss, Madsen, Oerlikon and Solothurn being manufacturers of these weapons. The Oerlikon 20 mm cannon was a 'blowback' type and was produced under licence in Germany as the MG.FF; this had a firing rate of 540 rounds per minute with a muzzle velocity of 2296 ft/sec. The Oerlikon had also been purchased by Hispano-Suiza in France, but the company were not satisfied with the weapon's operation and proceeded to design their own cannon in 20 and 23 mm calibre, which were gas operated with inertia locking.

The Air Staff of the RAF were interested in the Hispano-Suiza cannon (the air armament development in the United Kingdom having deteriorated badly since the end of World War I) and in 1936 representatives of the Air Staff and Air Ministry Gun Section visited Hispano-Suiza in Paris for a demonstration of the cannon. Although they were more interested in the 23 mm version, this could not be licence-built as it was under development purely for the Armée de l'Air. At the time the licensee in the United Kingdom for Hispano-Suiza engines and cannon, was Aero Engines of Bristol, so an approach was made to them to purchase eight 20 mm Type 404 cannon from the parent company. This was duly arranged, but due to the demand for the cannon, delivery of the first two was in two months, and the other six in six months. This cannon weighed 109 lb, had a firing rate of 650 rounds per minute and a muzzle velocity of 2880 ft/sec, as well as

an effective range of 2000 ft, so was more effective than the MG.FF.

The Hispano-Suiza cannon was termed a *moteur-canon*, as it was installed as part of the engine installation, with the gun nestling in the 'V' of the engine and the gun barrel protruding through the propeller shaft, using the mass of the engine to absorb the gun recoil through the gun mounting. To investigate this cannon and the engine installation the Air Ministry purchased a Dewoitine 510 fighter from France, which was fitted with the Hispano-Suiza engine and cannon installation; and although the airframe itself was obsolete, the cannon installation and also the gunsight interested the Air Staff, so the aircraft was tested at A&AEE Martlesham Heath.

The gunsight as well as the cannon installation also interested the experts at A&AEE, the gunsight being of the reflector type. Only ball-type ammunition was available and this was used during the tests of the cannon. The first

Dewoitine type 510 purchased by the Air Ministry in 1937 for investigation of the Hispano-Suiza 20 mm cannon installation. (RAE)

production aircraft installation of the cannon on a British aircraft, was on the Westland Whirlwind, when four HS 20 mm cannon were installed in the nose (the supervision of this installation was carried out at Boscombe Down* by Squadron Leader J. Munro).

The gun at this stage in its development was magazine-fed, the magazine holding 60 rounds; but at Chatellerault Arsenal a belt feed was under development but had not been perfected. The Bristol Aeroplane Company would also commence development of their Mk.1 belt feed almost in line with the development of the Beaufighter, and this belt feed used the gun recoil to operate the feed mechanism.

During the original weapon testing, the Colt 0.5 in machine-gun and 37 mm cannon were under investigation, but the 0.5 in Colt was rejected due to lack of development and a reputed barrel life of 7000/8000 rounds fired; whilst the 37 mm cannon was eventually air-tested in Whirlwind prototype L6844. During 1940–41 even heavier weapons would be air-tested, and these would be test-flown on the Beaufighter, which included a modified Bofors gun as well as the Vickers 'S' and Rolls-Royce 40 mm cannon (*see* Chapter 2: Increased Armament).

With the acceptance of the Hispano-Suiza 20 mm cannon by the Air Ministry as an RAF weapon, the Hispano-Suiza company set up a manufacturing base at Grantham under the title of British Manufacturing and Research Company Ltd (BMARCO), and the Chatellerault Arsenal began further development in line with British standards and modifications, with improvements of their own. The Royal Ordnance Factory at Poole were programmed to carry out any modifications on the 20 mm cannon, and after the fall of France would become the centre of research and development of this weapon, but that is not only another story, but three years further on.

The Cannon Fighter

The Hawker F36/34 and Supermarine F37/34 designs were by now revised to accept eight Browning 0.303 in machine-guns in their wings, but the Air Staff were in the process of upgrading their aircraft armament requirements, for it was known that a number of Continental aircraft were carrying 20 mm cannon.

With this upgrading it was necessary for the Operational Requirements Committee to call a conference to discuss these requirements in detail. This was convened on 29 March 1935, and had before it the basic requirements of two heavily-armed types of landplane fighter projects, namely:

(a) Two-seat day and night fighter to F9/35.

(b) Single-seat day and night fighter to F10/35.

These projects were to have an estimated performance as in Table 1, and to have an armament of not less than six cannon, but preferably eight.

After an inspection had been carried out of both the Hawker F36/34 and Supermarine F37/34 mock-ups, Squadron Leader R. Sorley minuted the ACAS to the effect that these two aircraft appeared to have forestalled the F10/35 specification, except as regards the armament, which was not in line with their present thinking. Quite an amount of thinking on similar lines was apparently going on within other departments, for on 21 May 1935 DTD R. Verney minuted the ACAS that the Air Staff might consider three alternatives:

(a) To let F10/35 continue

(b) To withdraw F10/35 for a time

(c) Recast the specification and make demands for an even more advanced design

In his minute he advanced his opinion that course (c) would be preferable, as some manufacturers might have already commenced work against the specification requirements, and so an early decision was required. This resulted in a letter being sent to all manufacturers on 21 June, notifying them that Specification F10/35 was to be suspended for the time being. Following this, F10/35 was recast as F37/35, calling for a single-seat, single-engined day and night fighter.

A number of firms tendered to the specification, amongst whom were Bristol, Supermarine and Westland. Bristol submitted their Type 153, which was a single-seat, low-wing monoplane derived from their Type 151 submission to F35/35. The Type 153 had the following data:

---

* This was prior to the A&AEE's move to Boscombe Down and was carried out on the Whirlwind prototype.

Estimated maximum speed	357 mph at 12,500 ft
Estimated service ceiling	33,200 ft
Estimated initial rate of climb	3580 ft/min
Armament	four 20 mm cannon
Engine	Bristol Hercules
Wing area	204 sq ft

Squadron Leader J. Munro in front of a Whirlwind, for which he helped to develop the cannon installation, and who helped development of the 40 mm cannon on the Beaufighter when OC AGME. (J. Munro)

Just afterwards a further issue of the specification was made, this changed the heading from 'Single-seat single-engine . . .' to 'Single-seat day and night fighter', and allowed the use of two engines.

Barnwell and Frise at Bristol meanwhile were in some doubt as to the wisdom of mounting guns of such heavy calibre outboard in the wings, both from the accommodation point of view and because of stress loading; so against the revised specification they submitted their Type 153A. This was a twin-engined single-seat low-wing monoplane powered by Bristol Aquila engines, having a slim fuselage which tapered to the leading edge of the mainplane at the front and had a twin-finned tailplane sitting on the top at the rear (Figure 1). In 1937 a USA patent was granted to Donove of Curtiss-Wright for an aircraft of similar layout. It was estimated that the top speed of the Type 153A with four 20 mm cannon mounted at the front of the fuselage would be 370 mph at 15,000 ft.

Out of all the projects tendered to the F37/35 specification only the Westland project was chosen to proceed to production status, this became the Whirlwind*; the other projects remained as paper designs. The Whirlwind was an advanced design with a number of novel features, but was unfortunately powered by the Rolls-Royce Peregrine engine, a further development of the Kestrel engine. The engine was at that time unreliable and under-developed, and would remain so as Rolls-Royce were concentrating their efforts on the Merlin engine. The situation was further complicated when the Air Ministry failed to place a production order, at a time when other aircraft were being 'ordered off the drawing board', then further delayed pending trials with the second prototype; so that the Whirlwind failed to enter service with the squadrons until late 1940, and then not in sufficient numbers.

In 1937 Frise and Fedden proposed a thick-winged version of the Type 153A, powered by two horizontally-opposed sleeve-valve engines, that were submerged in the main aerofoil, and covered by patent 482135. Bristol also tendered two other specifications, a twin-engined (Hercules) cannon fighter with a dorsal turret to F11/37, and a Centaurus-powered cannon fighter to F18/37, but neither tender was accepted. So the Munich crisis found the RAF without an escort fighter, a cannon fighter or a true night fighter; in fact, it was not until October 1938 that the first prototype Whirlwind flew, and no production order was placed until 1939. Even then, any aircraft without AI radar could hardly be classed as an efficient true night fighter.

At the Bristol Aeroplane Company at this date the airframe team was led by F. Barnwell, with Leslie George Frise as his right-hand man; Frise had joined Barnwell at Bristol in 1916 and would take over the design team on Barnwell's death in 1938. The engine design team was composed mainly of Alfred Hubert Roy Fedden, L.G. (Bunny) Butler and Fred Whitehead. Butler transformed Fedden's visions and ideas into practical designs, whilst Whitehead developed the designs to production status — Whitehead being an outstanding production engineer. Fedden was an engineering visionary, never satisfied, who upset many people with his attitude. He started as an apprentice with the Bristol Motor Company, then joined Brazil Straker as a junior draughtsman. By 1918 he was the company's chief engineer with Butler and Whitehead in the team. The team stayed together when the company became Cosmos Engineering, and became the Aero Engine Department when the British & Colonial Aeroplane Co. (Bristol Aeroplane Company) bought out the company.

With the lack of a cannon fighter, by 1938 the Air Staff realized that some stopgap aircraft would have to be produced, which initially

---

* *Whirlwind* by Victor F. Bingham (Airlife Publishing Ltd).

	F10/35 single-seat fighter			F9/35 two-seater fighter		
Specified performance 15,000 ft normal rpm. Hours.	1½ @ 267 mph	1¼ @ 269 mph	1 @ 271 mph	1½ @ 253 mph	1¼ @ 254¼ mph	1 @ 256 mph
Approx. endurance, 15,000 ft full throttle rate boost. Hours.	0.82 @ 310 mph	0.68 @ 312½ mph	0.53 @ 315 mph	0.82 @ 294 mph	0.68 @ 296 mph	0.53 @ 298 mph
Approx. endurance 15,000 ft economic speed. Hours.	4 @ 150 mph	3.5 @ 150 mph	3.0 @ 150 mph	3.6 @ 150 mph	3.1 @ 150 mph	2.7 @ 150 mph
Maximum speed, 15,000 ft	310 mph	312½ mph	315 mph	294 mph	296 mph	298 mph
Time to 15,000 ft. Minutes	4¼	4.63	4.5	5.5	5.37	5.25
Service ceiling. Feet.	35,000	35,300	35,600	33,000	33,250	33,500
Landing and take-off over 50-ft barrier	less than 500 yards					
All up weight. lbs	5350	5205	5060	5930	5782	5635

Table 1 — Estimated performance for two specifications

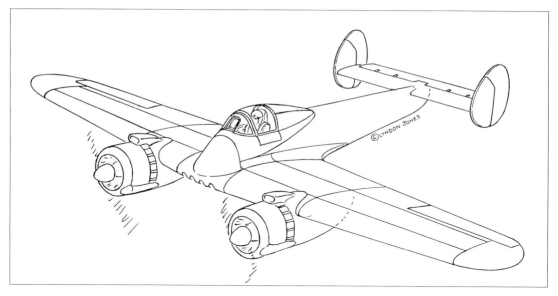

Figure 1. Artist's impression of the Bristol type 153A, Bristol's project to the F37/35 specification.

resulted in a proposal for the conversion of some Blenheim aircraft for that role; but this was not proceeded with. Air Interception (AI) radar originated exclusively in Great Britain, although radar had been developed by many countries AI radar made its flight debut in a Fairey Battle, and in July 1939 the AI radar-equipped Battle was flight tested with the AOC-in-C Fighter Command, Hugh Dowding, on board. The result of the flight test convinced Dowding of the necessity of AI radar as a further instrument in the defence of Great Britain — a further link to the CH Chain. His other conclusion was that AI radar should be fitted in a twin-engined aircraft for safety at night and with a separate crew member to operate it. It took until September 1940 before a number of Blenheim Mk. 1Fs were equipped with AI Mk. 3 sets, but the Mk. 3 radar was not an outstanding set and the Blenheim had too low an interception speed to overtake some of the enemy intruders.

The Blenheim bomber was a military redesign of the Bristol Type 142 light transport, but the relationship was not as close as that between the Beaufort and Beaufighter. The Blenheim was not designed to a specification, but was the result of the Air Ministry's interest in the Type 142, which encouraged Frank Barnwell to prepare a layout and performance estimate of the military 142. Yet the Air Ministry at the start of 1935 had no specification for a replacement of the Fairey

Battle light bomber, nor for a medium bomber like the Blenheim.

As though this had awakened the Air Staff, in September 1935 two specifications were issued, M15/35 for a land-based twin-engined torpedo bomber, and G24/35 for a twin-engined general reconnaissance aircraft. As both specifications could be met by an aircraft derived from the Blenheim, and it appeared possible that Bristol could design an aircraft to combine both roles, their design staff prepared a proposal, and this was submitted in April 1936.

This was designated the Type 152 by the company, and the Air Staff were prepared to accept the design with certain alterations; this entailed enlarging the fuselage to provide a navigating position forward, and raising the pilot's position. To cover the new design Specification 10/36 was issued. The main constructional difference between the Blenheim and Beaufort, as the Type 152 would be named, was that the Beaufort was of unitary construction.

On 2 August 1938 the Bristol Aeroplane Company lost the services of their chief designer, Frank Barnwell, who was killed in an aircraft accident, at a time when the company were fully engaged in the production of Blenheim and Beaufort aircraft. The Air Staff at this period had a problem, as already stated: no cannon fighter, no night fighter. This difficulty had already been

Bristol Beaufort torpedo bomber, the basis for the Beaufighter project. (Bristol Aeroplane Co.)

sensed within the industry, and Bristol's considered that they had the basis for conversion to a cannon-armed fighter.

The design team was now under the leadership of Leslie G. Frise, and the basis for conversion was the Type 152 Beaufort, and as the construction was unitary this made the job that much easier. On the basis of this, Frise submitted to the Air Ministry in October 1938, a brochure entitled 'Two-seater cannon fighter utilising components of the Beaufort based on Air Ministry specification F11/37'. On 29 November the Air Council met to consider the proposal and the effect that it would have on other aircraft production. Then during December the Air Staff decided that Fighter Command must rearm with cannon-armed fighters, and was already pushing both Hawker and Supermarine to upgun their single-seat fighters.

Hawker's answer was to sling a 20 mm cannon in a bulky cowling below each mainplane on Hurricane L1750. Flying tests at Martlesham in comparison with a standard Hurricane found it handling as well, and only 8 mph slower. Supermarine managed to lever one 20 mm cannon in each mainplane of a Spitfire. The 20 mm cannon was having its problems with both

the feed and the cartridge ejection, and on the Spitfire this was compounded by Supermarine installing the cannon on their sides. The Spitfire's mounting for the cannon in the mainplanes was not as rigid as that on the Dewoitine, and with the cannon mounted on its side, this meant that the bolt ran along one slide in its operation, which was not the cannon's correct operating position.

The ACAS, W. Sholto Douglas, on 29 November 1938, wrote to the CAS about Bristol's proposed conversion of the Beaufort. He mentioned that of the various proposals examined, that submitted by Bristol was the most promising, and: 'Their suggestion is that the Beaufort should be used as a basis for a new cannon fighter. They originally proposed mounting two Hispano guns in a turret, but they also submitted an alternative proposal for four fixed Hispano guns. D/DOR and I have examined these proposals, and we are of the opinion that the scheme for converting the Beaufort into a fixed-gun fighter is the better of the two.'

Air Marshal Sir Wilfred Freeman asked Bristol for further details of their project, which were presented to him on 23 December. More discussions took place on 2 January 1939 at the Air Council, when it was brought to their attention that Bristol's had already commenced serious design work, had cannon installation experience and were at that time involved with work on a twin-cannon Blenheim for a foreign power.

The Bristol Beaufort fighter project as visualized by the design team would have the Beaufort and Beaufort-fighter main components (wings, rear fuselage and the tail unit) produced together on the same jigs, allowing a quick change of assembly to whichever aircraft had priority. The wing strength of the Beaufort was more than adequate for the fighter project as foreseen, and with the shortened nose of the fuselage to counterbalance the weight of the Hercules engines, it was determined that the tail unit would be adequate. As the Bristol Aeroplane Co. was a large quantity manufacturer, and its reputation one of reliability, the company was — as opposed to Westland — viewed with favour over the 'Beau' fighter project. There were a few doubts expressed over the size of the aircraft for a fighter, but the point of no return had been reached and no other manufacturer appeared to be able to produce such a type of aircraft quickly.

On 23 December Bristol presented Air Marshal Sir W. Freeman with further details of their project and promised that their Beaufort fighter would be ready for testing within six to eight months if an order was placed. A statement of the operational requirements to Bristol had the project hardening into a twin Hercules powered, two-seat fighter armed with four 20 mm Hispano-Suiza cannon. After consideration, the Air Staff felt that the project's size would restrict its manoeuvrability, but approval of the project was given. In early January 1939 it was agreed not to put the specification out to tender so as to speed production, and so close the gap in the RAF's defences. At this stage it was the Air Ministry's intention that the engines for this fighter project should be either the Hercules or Rolls-Royce Griffon, but due to an over-optimistic evaluation of the Hercules production programme, the Griffon was allocated to the Royal Naval aircraft programme.

Although the Specification F17/39 was not finally approved until July 1939, the early stages of normal design and development procedures were omitted and the requirements and limits of design were discussed verbally at an Advisory Design Conference, also decisions were made on the spot during the examinations of the mock-up.

This procedure, or lack of normal procedure, resulted in a marked reduction in the length of the development period of the Beaufighter; yet the method of control adopted allowed the Air Ministry to adequately control the design. Great credit reflects on both Bristol's and the Air Ministry, for the initiative in the firm's proposals and the Air Ministry's readiness to abandon established procedures. Draft operational requirements were forwarded to the firm prior to the mock-up conference on 11 March 1939, and guidance given in design matters; the requirements included the following important points:–

(a) Hercules VI engines
(b) Speed not less than 350 mph at 15,000 ft
(c) Four 20 mm Hispano cannon

That there would be delays in production of the Hercules VI and problems in the redesigning of the airframe was not totally unforeseen, but the Air Ministry's decision to press on with the prototype prior to the issue of the specification resulted in a shorter development period. There were however growing doubts that the Hercules VI engine production estimates were over-optimistic, and it was on these that the estimated performance of the 'Beau' fighter was based, and the deterioration in this performance and the problems encountered are covered in Chapter 2. The Hercules VI was the Hercules III modified to accept 100 octane fuel instead of 87 octane, and a higher supercharger gear ratio.

At the beginning of 1939 the forecast generally was that the Beaufighter would follow the Beaufort into production early in 1940, with the prototypes and hand-built pre-production aircraft being ready in 1939. Aircraft production staff estimated that there would be a delay in Beaufighters powered by Hercules VI engines, and that there would be about thirty Beaufighters powered by Hercules II engines. At a meeting on 20 June 1939 the Air Staff had to accept the unwelcome fact that a certain number of Beaufighters would have to be engined with Hercules IIIs, as well as the batch with Hercules IIs.

The Bristol design team's desire to use the same jigs for both the Beaufort and Beaufighter, and the need to use the bigger and more powerful Hercules engine, resulted in a number of changes that introduced other problems, though not insurmountable. The greater engine power required larger diameter propellers; so to accommodate this and ensure adequate ground clearance, the engine nacelles were repositioned from the under-slung Beaufort position to the mid-position. The nacelles used initially were Beaufort ones, as was the undercarriage. During the initial design and development period it was recognized that allowance must be made for increase in aircraft weight, so the Beaufort's oleo legs were replaced fairly early on with Lockheed long-stroke units. To ensure adequate propeller clearance with the long-stroke oleos, the propeller diameter was then reduced from 13 ft to 12 ft 9 in.

With the delay in delivery of the Hercules III and VI, Rowe (AD/RDL) wrote to Frise on 6 November 1939 suggesting that an alternative engine should be considered. The Double Cyclone and Griffon were two engines mentioned, but the Griffon was a doubtful contender as it was already on the FAA programme as well as posing a large departure from the projected layout of the Beaufighter. Answering this, Frise pointed out that the performance with the Double Cyclone would fall short of that with the Hercules, and on 23 January 1940 the DGRD agreed that consideration of this installation be dropped.

In the original proposal, the secondary armament of eight 0.303 inch Browning machine-guns was to be housed in a blister under the fuselage to the rear of the cannon. Doubts about this arrangement was expressed by the ACAS, who felt that the machine-guns should be installed in the wings. This was communicated to the RTO at Bristol, but Frise pointed out that incorporation of four machine-guns in the port wing would cause a delay in delivery of the prototype, as it would be necessary to reposition the landing light. Frise made an alternative suggestion to ease this situation; to place two machine-guns in the port wing and four in the starboard wing. The DGRD agreed to this suggestion on 15 March, and so the layout became part of the standard armament.

By this date the Beaufighter mock-up had been viewed by Petter, technical director of Westlands, who considered that the fuselage was too large, to which Frise replied that in two years time it would not be big enough — this was to be proved correct. On 23 May the AMDP notified Sir Stanley White of Bristol's that the Beaufighter was first on priority after the original first five priority aircraft. A further mock-up was viewed in 1940; this was the long-range fighter version for Coastal Command, which included a mock-up of the W/T and navigation equipment. So already Frise's prophesy was proving correct.

During this period radar development at Bawdsey continued, and a Radar Flight was operated by Martlesham on behalf of Bawdsey; the first twin-engined aircraft on this duty being Blenheims K7033 and K7034. These were fitted with AI Mk. III radar, which had transmitter power up to 5 Kw; this AI gave healthy signals on a target at a range of three to four miles. The first Beaufighter to receive AI equipment was R2055, which was delivered to the Fighter Interception Unit on 12 August 1940 complete with AI Mk. IV.

Chapter 2
Beaufighter Development

Conversion and development

Bristol in their brochure had estimated that their Beaufort fighter would have a maximum speed of approximately 360 mph at 15,000 ft at a maximum take-off weight of 15,500 lb; as will be seen later in this text this was wildly optimistic. Nevertheless, Bristol began the construction of the converted production Beaufort airframe immediately, having promised the Air Ministry that the Beaufort fighter prototype would be ready in nine months; the firm also considered that 100 fighters could be produced by 31 December 1939. On 7 February 1939, the Air Council decided that Bristol should drop their original planned production of Short Stirling bombers and commence production of the Beaufort fighter, and on the same date authorized the order of 300 of the aircraft from Bristol. This was confirmed on 24 February with the requisition order for this number of aircraft at a cost of £11,000 each, the requisition including the two prototypes (R2052 and R2053) utilizing components from the 71st and 72nd Beauforts.

At the beginning of March the firm decided to armour the prototype aircraft to their own layout, as no other requirements had been laid down; their layout was to provide non-magnetic armour to protect the instrument panel; vertical plate behind the nose piece; windscreen of bullet-proof glass; magnetic armour plate aft of the rear spar and protection of the fuel tanks. On the 31st of the month at the Preliminary Mock-up Conference a number of alterations were called for; these included changes to the armour plate, repositioning of the rudder and aileron trimmers; the GM2 gunsight to be mounted so that it could be easily detached in flight, and the provision of a ring and bead sight as an alternative. Fedden had hoped to provide HE6SM engines using 100 octane fuel for both the Stirling and the fighter, but this development fell well behind schedule;

this provoked a letter from G.T. Bulman (DD/RD Engines) to the DTD, in which he gave as his considered opinion that Bristol Engines' enthusiasm had been allowed to outstrip the discretion of experience, and that the possibility of the HE6SM with higher supercharger gear ratio and higher grade of fuel was obviously going to take longer than estimated.

The construction of the prototypes had progressed smoothly at first, but this was soon interrupted, for the possibility of taking a Beaufort airframe from the production line and converting it as had been expected, turned out to be a more formidable task. The layout of the fuselage of the Beaufort torpedo-bomber and the Beau' fighter was so dissimilar that this solution proved impracticable, and the whole fuselage had to be redesigned. The conversion of the Beaufort to Beau' fighter took 2,100 drawings, so that with both types of aircraft going into production at the same time, this extra drawing office work resulted in delays to production. A further handicap that ensued was the lack of a definitive specification and the firm was unable to obtain from the RDQ officers the internal layout for the type.

By the end of April 1939 Bristol Engines had reached the unpleasant decision that they had to inform the Air Ministry that their production forecasts for the developed Hercules VI engines had been optimistic, and this meant that a number of production Beau' fighters would have to be powered with the Hercules III engines. Then in June this position deteriorated even more, for it became apparent that the Hercules III would not be in production early enough, and that a small number of Beau' fighters would have to be equipped with Hercules II engines. At the Air Ministry the engine staff in reviewing the information came to the conclusion that Hercules III engines would not be available until February

1940. Then came the problem of propellers, for although Bristol's preferred to fit Rotol fully-feathering propellers, Rotol were not in production in time to meet this demand; in any case, the first batch of Hercules II engines destined for the Beau' fighter had no provision to accommodate fully-feathering propellers, and so the standard DH constant speed type would have to be fitted.

The Specification F17/39 and draft Appendix B were sent to Bristol on 17 May, and on the following day the AD/RDL received information on the Beau' fighter powered with the Hercules III and HE6SM engines (Table 2). This table indicated that with the Hercules III engines installed the maximum speed at 15,000 ft would be an estimated 336 mph; when this was fed through to the DOR, the Air Staff were prepared to acquiesce on the Hercules engine production position, but the Hercules VI installation on the aircraft remained their aim, and the DOR refused to allow the reduced speeds with Hercules III engines to be quoted in the specification. On 26 June 1939 the first production order was placed,

followed on 3 July with the approval of the specification by the DTD; the contract covered airframes, construction numbers 9562 to 9861, whilst the heading on the specification read: 'Fixed gun fighter development of the Beaufort G.R design', and in the preamble it stated '. . . development of a suitable existing design as an interim type to precede the normal replacement for the F37/35 design'. Prior to this, and even after this, on the strength of the brochure and initial communications, a number of Ministry personnel were under the impression that the fighter project was a direct conversion of the Beaufort, and not two distinct types, the Beau' fighter and the Beaufort G.R. So by the time that the draft specification had been issued, the RTO at Bristol was requesting that the title be amended from 'Fighter conversion of Beaufort' to 'Beaufighter Mk. 1'. This was further clarified by a letter dated 17 May 1939 from F. Cook of DTD to the Chief Superintendent RAE Farnborough, that stated:

> This aeroplane is a design conversion from Beaufort 10/36, but is not a conversion in the sense that a complete Beaufort can be modified into the form of a fighter.

So the name Beaufighter was adopted in all further documentation and contracts.

First prototype Beaufighter R2052 with oil coolers under the engines and undercarriage doors that do not enclose the wheels completely.

	Hercules III	Hercules HE6SM
Maximum bhp for take-off	1330	1550
Maximum bhp for level	1400 at 5000 ft	1610 at 5000 ft
	1290 at 14,000 ft	1510 at 15,000 ft
International rating	1150 at 5000 ft	1400 at 5000 ft
	1075 at 12,500 ft	1275 at 15,000 ft
Maximum speed at S.L	297 mph	313 mph
Maximum speed 10,000 ft	320 mph	338 mph
Maximum speed 15,000 ft	336 mph	361 mph
Maximum speed 20,000 ft	333 mph	356 mph
Rate of climb at S.L	2310 ft/min	3040 ft/min
Rate of climb 10,000 ft	2080 ft/min	2600 ft/min
Rate of climb 15,000 ft	1890 ft/min	2680 ft/min
Rate of climb 20,000 ft	1405 ft/min	2120 ft/min
Rate of climb 30,000 ft	430 ft/min	1000 ft/min

Table 2. Bristol estimated performance with Hercules engines at all-up weight of 16,500 lb

Third prototype Beaufighter R2054. Oil coolers repositioned in wing leading edge but undercarriage doors as for R2052.

In spite of the prototype being slightly delayed in its manufacture and conversion, and the delay encountered due to the unexpected drawing office work, the prototype (R2052) was completed by 20 July 1939, and made its first flight on that date.

One requirement of the F17/39 specification called for an approved type of cartridge starter with hand-turning gear to be provided for the engines, and over a period of time there was disagreement and discussion over this between Bristol and the Ministry. Bristol's wished to fit an electric starter on the engines, but the Ministry wanted the Coffman cartridge starter; as there were no Coffman starters with hand-turning gear it was hardly relevant for the Ministry to keep stating that this was what was required under the requirements. Hope was expressed by the Ministry that Coffman's development of a starter with hand-turning gear would be available in six to nine months' time.

Previous to this, on 3 May 1939 Bristol's had asked for a concession regarding the first two Beaufighter prototypes, asking that they would be accepted with the guns tilted upwards at an angle of 5° instead of horizontal, as would be the case with subsequent aircraft; the concession was granted on the 19th of the month. During May the preliminary draft specification was being discussed at Bristol between Ministry representatives and the company, during which the company stated that the existing aircraft (Beaufort) complied with the strength factors specified. This was queried on 9 June in a letter from RAE Farnborough to the Secretary of State for Air, when in referring to the structural strength, it was pointed out that this was in general, that usually required by a medium bomber and therefore too low for a fighter. It was also pointed out that the terminology in the specification stating a requirement of a fast *glide* at 400 mph would be better specified as terminal velocity attitude. This requirement was quickly satisfied, when Bristol's stated that the airframe was satisfactory for a terminal velocity dive of 450 mph, so the 'fast glide' condition was increased from 400 to 450 mph.

Strangely enough, even in October 1939, after the first flight of the first prototype and the issue of a production contract to Bristol's for 300 Beaufighters, still no production specification had been issued. Early on in the month Treasury approval was sought for the acquisition of no less than 1300 Beaufighters, the production being provisionally split up as follows: 500 from Bristol (including the 300 already ordered), 500 from Fairey and 300 from Supermarine's Shadow factory. This contracting was altered on 14 December 1939, when the Air Council decided that part of the Fairey manufacturing capacity would be reserved for production of Beaufighter spares on the basis of 15 per cent of complete aircraft produced.

On 29 November the programme of prototype aircraft was laid out; this agreed that the first and second prototype should be regarded as unrepresentative aircraft, as neither the armament nor the windscreen was correct. By the end of October the RTO at Bristol's wrote to the DAD/RDA, that the Beaufighters constructed to the production contract were only covered by the Specification F17/39, and that the first prototype was still flying without equipment, and the second one would soon be completed in the same condition. As programmed, the first prototype was to be used on general aerodynamic handling trials flown by A&AEE pilots, and the second prototype to be used for engine cooling trials at the works.

Contractor's trials and test flights occupied the next six months after the first flight of the prototype; this period allowed the usual teething problems to be eradicated, as well as a number of modifications shown up by these problems to be incorporated. So the prototype Beau did not get delivered to A&AEE until February 1940, though it had already been flown by A&AEE pilots at Bristol's Filton base in January, to gain experience on the type, as well as preparing a preliminary test report. The possibility of a dual control Beaufighter was on the agenda during the middle of 1939, but Bristol's Chief Test Pilot, C.F. Uwins, wrote on 23 October to Farren (DD/RD), giving his reasons why he considered that no dual control version was necessary. In his opinion the Beaufighter was easy to fly, had great reserve of power, and that any pilot with Blenheim or Beaufort flying experience should have no difficulty in flying the Beaufighter. Obviously this advice was taken, for no dual control Beaufighter was ordered or produced.

When the prototype first flew, the oil coolers were installed below the engines, which were Hercules HE1SMs with de Havilland two-pitch type propellers of 13 ft diameter, though Bristol's were still proposing Rotol fully-feathering constant-speed propellers for production aircraft — the second prototype was to have similar propellers also. Handling was in general satisfactory, but longitudinal instability on the climb was present, whilst the reduced manoeuvrability mitigated against the type's use as an escort fighter.

In the 'clean' state R2052 achieved its estimated maximum speed of 335 mph at 16,800 ft. Minor changes were required to improve the aircraft's conformity to the operational requirements; the elevator control circuit was stiffened, fin area increased and the main undercarriage revised. The two prototypes were delivered in February 1940, and a final conference was held on 19 February to discuss them, and to determine the alterations and modifications to be incorporated into production aircraft. The other prototypes were hand-built — these were listed as pre-production aircraft — and resulted in the third to sixth aircraft being modified up to the requirement standard agreed upon, and the planning for installation of the Hercules III engine. All this, plus the engine hold-ups, resulted in delivery of the hand-built aircraft being delayed until May 1940.

On 2 May the Final Inspection Conference on the Beaufighter armament installation was held, and covered a continuous-feed cannon installation on R2053, a drum-feed installation on R2054, the stowage of spare drum feeds and the method of firing the Browning machine-guns and cannon. At the time of the inspection all the guns were fired simultaneously by a single button; DDOR however considered that separate firing of the machine-guns should be made a future requirement, and it was arranged for Dunlop to develop such a device.

Allocations were made on 8 May for the first eleven Beaufighters, R2052 to R2062 inclusive, for test purposes; but due to the demand for the aircraft in service, within the month four of these aircraft that had been allocated to AMDP for development work, were then diverted for issue to service units; five maintenance units at this date having been named for Beaufighter storage.

On 4 June a request followed from N.E. Rowe (AD/RDL) to the RTO Bristol, asking for confirmation that R2052 and R2053 could be fully equipped for operational use (non-standard) within three weeks. The rush to get the Beaufighter into squadron use was speeding up, for on the 13th the draft flying limitations were being issued for the Beaufighter Mk. I with Hercules III engines; and although the firm's test pilots had dived the aircraft to 400 mph, the limitations advised caution in diving beyond 360 mph — the limit of the speed dived at A&AEE.

Pre-production prototype R2055 at Filton 1940. Undersurface of mainplane one half black, other half white.

A liaison meeting was held on 7 August 1940 between the DGRD and ACAS(T), and the subject was airframes and engines. The performance of the aircraft with Hercules and Merlin XX engines and the redesign of the fuselage were the main topics. It was agreed that the redesigned Beaufighter with the new slim fuselage (Type 158), when powered with Hercules III engines could be the Mk. III, and when powered by Merlin XX engines would be the Mk. IV. It was stated that due to fundamental changes in dimensions there would be little prospect of using the Mk. I and II fuselage jigs, but the work was to be on high priority. The chances of this coming to pass became nil with Beaverbrook's list of priority aircraft, the end of the Battle of Britain and the start of the night Blitz.

Returning to 1939 and the origins of the Beaufighter, it was envisaged that only the forward fuselage and engines would be replaced, and Frise considered that by raising the engine nacelles the same undercarriage could be used. It was thought at that time that 75 per cent of the Beaufort jigs and tools could also be used, and that although the fuselage was smaller with a shortened nose, it would be possible to arrange the same jigging attachment points on the centre plane. As the work progressed many other changes were required, and although only 2,100 drawings were required in the conversion from Beaufort to Beaufighter, twice that number were required to get the Beaufighter operational.

In January 1939 it had been determined that certain members of the aircraft structure required strengthening to meet larger loads; these were the centre-section rear spar web, aileron spar tube, engine nacelle tubes, tailplane spar and elevator spar. In order to modify the parts without drilling, riveting or assembly, the material was changed or the metal thickness increased. In this way there was no change to the jigs or sub-assembly manufacture, and thus no hindrance to the production line. Jigging and tooling proceeded parallel with that of the Beaufort, but then the general use of common parts began to become less than had been expected. It was found that the whole fuselage required redesigning — this and the teething problems of both the Beaufort and Beaufighter, put an excessive workload on the Drawing Office staff, which resulted in a delay with

dealing with urgent aspects of both designs. Delays were also encountered with the jigging and tooling for the undercarriage attachments and structure; which resulted in there being variations between various individual Beaufighter aircraft; some of this was blamed by the Ministry on bad co-operation between the design office and experimental department.

When the third prototype R2054 in full operational trim was tested at A&AEE Boscombe Down in June 1940, it returned a maximum speed of 309 mph at 15,000 ft, even though powered by two Hercules III engines. With this, a development programme was commenced by Bristol, which was intended to obtain improvements in performance and handling. One of the sources of drag was the nacelles, and this was improved by lengthening the nacelle over the top of the wing and reducing the cross-section below the wing. Undercarriage doors were made which completely enclosed the wheels; the outboard fuel jettison pipes inherited from the Beaufort were deleted and relocated in the tail of the nacelle; while the oil coolers were transferred from underneath the engines to the leading edge of the mainplanes, as on the Beaufort. This resulted in R2060 powered by Hercules III engines, returning a maximum speed of 323 mph at 14,400 ft, even though the undercarriage doors did not close properly.

Production-wise a number of problems were affecting the Beaufighter, as they were to other types of aircraft, mainly due to the lack of large production orders pre-war. One problem was highlighted by the Director of Aircraft Production at the first meeting of the Beaufighter Group Works Managers, when he stated that the 'first bottleneck' would be light alloy forgings, ferrous forgings and light alloy extrusions. Another problem affected Bristol direct, for the Radcliffe Tool Company of Park Royal, who were proposed to undertake the tooling for the Beaufighter spar production, were unable to undertake further work, and were proposing to set up a factory at Llandudno, but were still awaiting financial sanction from the Treasury. It was now 1940 and Beaufighters were wanted. The Bristol factory at Old Mixon, Weston-super-Mare, was approved; and two buildings for Beaufighter production were proposed, with production estimated to commence in August.

At the Materials Committee on 19 June 1940, it was confirmed that Boulton & Paul was no longer in the Group, as they had to boost Defiant production. Although the Group had considerable supplies of materials, there were certain shortages, and there was little evidence of this being altered.

With the apparent restricted production of the Hercules engine for the Beaufighter, it was suggested that both the Wright Cyclone and Rolls-Royce Griffon engines should be considered for installation. Frise was however of the opinion that the double-row Cyclone engine did not offer any improvement in performance, whilst the Griffon was retarded at Rolls-Royce due to their priority development of the Merlin engine. The DGRD agreed with Frise on this point, and the idea of using either of these two engines was dropped, with the Merlin as an alternative being proceeded with. Three airframes were selected for this conversion, and it was planned that these should be delivered to Rolls-Royce Hucknall in April, May and June 1940, with the whole conversion to be completed by July — this plan in hindsight appears to be unduly optimistic — and was not achieved anyway, as the airframes were not delivered on time.

Apart from the fact that a number of pre-production aircraft were allotted to various establishments for tests or trial installations, other work was undertaken at Filton. R2060 for instance was given a special treatment: the application of a smooth filler over all rivets and joints, the joints being taped over before the application of the filler. In this condition the aircraft was flight tested, and the maximum speed was raised to 329 mph at 16,000 ft, which appears to suggest that the standard production finish was good. A further test that was carried out at Filton concerned the fuselage of the seventh airframe; so as to have a comparison of strength and drag the construction was carried out with mushroom-headed rivets — the standard riveting method using flush rivets. The results obviously satisfied the manufacturer as no change in riveting was made.

One fault that had shown up on the prototype during flight testing, was the opening of the undercarriage doors during flight manoeuvres (this would also occur with the Mk. II). By mid-1940 R2060 was on trials at Filton with a modification to prevent this, after which all subsequent aircraft were fitted with modification 287.

Two pre-production aircraft, R2054 and R2055, were delivered to A&AEE for preliminary trials on 4 June, but in early August R2055 was routed back to Filton for a check. This aircraft on 11 August was delivered to St Athans for special installation, prior to its collection by FIU (commanded at that time by Wing Commander G. Chamberlain) based at Tangmere. This aircraft was taken on the strength of FIU on 13 August, but three days later received minor damage during an air raid on Tangmere, and so was returned to Filton for rectification.

Approximately about this period the Beaufighter's clearance weight came up for review by the MAP and Air Ministry; for the original Beaufighter structure was designed to meet the full specification requirements with an all-up weight of 16,800 lb. With the all-up weight now increased to 18,250 lb, the structure failed to meet the requirements, so a number of modifications were introduced that either thickened material gauges or changed materials at a number of positions. By the end of June a trial Beaufighter had been flown and cleared in all forms of flight to an all-up weight of 18,500 lb; the RTO at Bristol's informing the RDL2(a) that the aircraft could undoubtedly be cleared to still greater weights when required.

During November 1940 R2066 was attached to the RAE Farnborough for tests; whilst on one of these, and flying at 12,000 ft, two Hurricanes made what was termed as 'threatening mock attacks'. The flight test observer scrambled forward to the signal pistol and fired off the colours of the day. The two Hurricanes then commenced to come in on a rear attack, and as the pilot's vision to the rear did not allow him to see what was happening, he put R2066 into a steep dive and escaped. As considerable time was lost in the observer getting to the signal pistol, it was recommended that a semi-automatic pilot-fired signal pistol should be installed to allow recognition signals to be fired off quickly when challenged. The Bristol Aeroplane Company had already commenced installing this type of pistol, so there was no problem in requesting the modification.

This was probably not the first, and certainly would not be the last attack on a Beaufighter by 'friendly' fighters, or during interception of 'friendly' bombers — faulty aircraft recognition in confusing the Beaufighter and Junkers 88 was the trouble, or to put it bluntly 'finger trouble'. One of the worst occurred on 23 December 1942, which had a delaying effect on the development of AI IX; this came about when TFU Beaufighter VI V8387 operating from Coltishall in conjunction with a 68 Squadron Beaufighter, were checking AI Mk. IX against 'Window'. Dr. D. Jackson was flying passenger on the 68 Squadron aircraft and Squadron Leader H. Mould and Dr. A. Downing crewed the TFU Beaufighter. At 1545 hours three miles off Kings Lynn, Norfolk, a 'friendly' Spitfire flown by a Canadian shot at both aircraft, shooting down the TFU aircraft and killing both crew; the 68 Squadron aircraft limped back to Coltishall.

Two pre-production Beaufighter Mk. Is, R2056 and R2057, were at the Air Fighting Development Unit at Northolt in September 1940; these aircraft as tested were fully loaded at 18,830 lb and were powered by Hercules III engines of 1300 hp each. Both aircraft had a modified elevator but unmodified ailerons. Their report issued on 11 September was quite comprehensive, including day and night flying and armament testing.

Regarding handling by day it said:

Aircraft found to handle as described in handbook, but the impression was gained that the stall is not so vicious as stated. Nevertheless, there is a feeling that the aircraft is a heavily-loaded one and care must be taken when near the ground to maintain full flying speed or the aircraft will drop out of your hands. The take-off is straightforward, although there is a slight tendency for the aircraft to swing to the starboard; this is easily corrected by use of the rudder. It is necessary to raise the tail quickly and to hold the aircraft down on take-off until the critical speed of 150–160 mph has been attained. In the air it is pleasant to fly. The lateral control is a little heavy, but it is understood that the modified ailerons greatly improve this feature.

Handling at night was described as:

The take-off is quite normal and the pilot's view in the air is good . . . It is most important that the throttle levers are so adjusted that when together, both engines are revolving at the same speed. The engines are rather silent and unless this is done, there is a possibility when landing of having unequal revolutions which results in yawing.

A number of cockpit fittings were criticized, many referred to the cockpit visibility, such as difficulty in reading compass, gill position indicators not readable at night, reflections of luminous dials on windscreen. The GM2 gunsight was considered too large, in that it obscured a large part of the pilot's vision, and so a GJ3 gunsight was fitted in lieu.

Tactical manoeuvrability was also investigated, and this found that:

Manoeuvrability at low altitudes is reasonable considering the size of the aircraft, and when the new ailerons are fitted it is anticipated that it will be quite good. However, it is thought that the aircraft will not be sufficiently manoeuvrable to participate in dogfights.

Having regard to the above, it still must have been daunting for a young wartime pilot, with little twin-engined experience, to arrive at an OTU, and find on arrival the unit equipped with Beaufighters; its pugnacious look and large powerful engines were outstanding. To be fair, in the early days the handling of the Beaufighter was considered tricky, due partly through lack of experienced Beaufighter instructors. The Beaufighter's two powerful Hercules engines with large propellers converted the power to a lot of thrust, and this propwash spiralled behind each engine and gave a large sideload on the fin. Combine this with the heavy rotating mass of the two power units and you have the makings of a swing!

With the Bristol Aeroplane Company incorporating the BLG type tailwheel unit onto the Blenheim V (Bisley), it was a follow-on to incorporate the same type of tailwheel unit onto the Beaufighter. This was a natural for this aircraft, for the all-up weight of the Beaufighter had been on the increase ever since its entry into service, and shimmy had become a problem with the Lockheed tailwheel unit. The Lockheed unit had been fitted with a modified oleo strut, a Marstrand twin-contact tyre, and for Coastal Command a tailwheel lock had been incorporated. The BLG tailwheel unit was of the trailing-link type and eliminated shimmy.

RAE Farnborough Structural and Mechanical Engineering Department were involved over a number of years in improving the undercarriage of the Beaufighters, as they were with other types of aircraft, but in the case of the Beaufighter this mainly affected the tailwheel unit. They cleared the BLG type on the Mk. X aircraft, with the proviso that twin-contact tyres were not to be used in conjunction with a tailwheel unit strut that incorporated an anti-shimmy device; this was the BLG type and the Lockheed AIR28554 tailwheel unit.

Armament was another point that came under discussion and planning, and the fact that it had no rear armament gave rise to a statement that the aircraft was not a viable escort fighter. A further point raised was regarding the fitment of a Boulton & Paul turret, but with such an installation lowering the Beaufighter's speed down to 300 mph, this also reduced its viability. The nacelles had no space either to accommodate fixed rear-firing guns, although this was another suggestion put forward, as were machine-guns in the wings. There was no need to discuss the possibility of the Beaufighter as an escort fighter, for as the AFDU report had pointed out, it was not manoeuvrable enough in this role.

With Coastal Command's requirement for a long-range strike-fighter at hand, Bristol converted R2152 to this role, and the aircraft was available for trials in late November 1940. Although the Command wanted extra armament for defence at the rear, in the form of either a rear defence gun or nacelle-mounted 'scare' guns, in the interests of early availability and for fear of slowing down production, this requirement was rejected.

The Beaufighter was at this period on operational trials and in production, and suffering the pangs of a restricted intensive flying programme, and the rush to get the aircraft in full production. This resulted in December 1940 in an appraisal and analysis of the Beaufighter programme being carried out by the Air Ministry, with a possible replacement of the aircraft by the Gloster F9/37, which would appear to have been favoured by some members of the Air Staff.

The Gloster F9/37 was derived from a design study for a twin-engined turret fighter to

Specification F34/35; this had never reached fruition, but the design was there. The first prototype of the F9/37 (L7999) with Bristol Taurus engines, reached on test a maximum speed of 360 mph at 15,000 ft; whilst the second prototype (L8002) was powered with the Rolls-Royce Peregrine, and reached a maximum speed of 330 mph at the same height.

However, after consideration of the two aircraft, which needed little arithmetic(!), it was realized that however much the Gloster F9/37's performance impressed, it was only at the start of its testing and development period, before it could go into production; whereas the Beaufighter was in production and being delivered. A further question that must have hung over any decision over the choice of the Gloster, was the location of the aircraft's armament; for the five 20 mm cannon were disposed two in the nose and three behind the pilot, all angled up at an angle of approximately 20 degrees. With such an installation, the re-arming would have been awkward, and the clearance of stoppages or changing ammunition drums an impossibility in combat.

However impressive the Gloster may have looked to some, the fact was that at the end of 1940 there were only two prototypes of the F9/37 available, whereas there were 111 Beaufighter airframes already at Maintenance Units or squadrons, with production planned at a Shadow Factory at Weston and by Fairey Aviation.

With regard to the production of Beaufighters, Air Marshal Sholto Douglas requested information on it from MAP. This, on 20 May 1941, prompted a reply from Westbrook of MAP, which stated that Bristol's were very weak and gave MAP endless trouble, that it was difficult to liven a place up that had been dead for years! Apart from being both unnecessary and uncomplimentary, it would appear to be a fatuous comment in view of Bristol's past response with the Blenheim and the Beaufighter to fill gaps in the RAF's armoury. Was it not Beaverbrook and the MAP, who in 1940 had placed the Beaufighter on a lesser priority than the Blenheim, Hurricane, Spitfire, Wellington and Whitley? In addition the RAF had dithered over ordering the Whirlwind, and had failed to see the need for a *real* night-fighter; now those flying mahogany desks at MAP were trying to blame the firm that had created the aircraft to fill the need that they and the Air Staff had not foreseen nor planned for.

In June comparative performance tests of the various Beaufighter marks were made by the A&AEE, in response to a request from the AMSO's department. Although the prototype Mk. II when tested at Hucknall gave a top speed of 350 mph, subsequent tests of the same aircraft at A&AEE failed to produce the same results. The comparative figures submitted by A&AEE were:

	Mk. 1 Hercules III	Mk. II Hercules XI	Mk. III Merlin XX
Maximum range	1620 mile	1620 mile	1560 mile
Maximum speed	330 mph	340 mph	337 mph
Ceiling at max. weight	25,000 ft	25,000 ft	30,000 ft
Maximum weight	21,000 lb	21,000 lb	21,000 lb

The cannon of the Beaufighter was at this point the weakest part of the design, and many squadron aircraft were unserviceable for this reason; so R2060 was sent to A&AEE for trials covering the 20 mm cannon and its ammunition feed, also the Bristol Mk. I belt feed. Difficulties were experienced with the feed chutes, ejection of the spent cartridge cases, and with the belt feed. Trials of the Mk. II cannon also found that its functioning was not as reliable as later production Mk. I guns.

A further test carried out during this period was commenced at the RAE Farnborough; this involved a standard Beaufighter Mk I nacelle fitted with a Hercules VI engine mounted on a wing section of 20 ft span. This was mounted in the 24-ft wind tunnel and tested with flared and unflared propellers; the two types tested were the 12 ft 8 in DH 6/6 type and the 12 ft 9 in Rotol 11/36/10, both types having three blades. The object of the tests was to determine the effect that the propeller blade root design had upon cowl entry losses and the swirl angle of the cooling air entering the cowl, and to measure the flow through the exhaust shroud. From these tests would be determined the optimum blade type, the propeller and whether to fit a spinner.*

A number of tests were carried out in the wind tunnels at RAE Farnborough, using models of the Beaufighter to determine the spinning characteristics. The first test on the Hercules-engined Beaufighter confirmed that the recovery from a sustained spin, by normal control movement, was not likely to be satisfactory. Such tests were of course complementary to actual flying trials carried out at RAE and A&AEE. Amongst the aircraft allotted for this latter work were R2052, R2054, R2055 and R2060.

The Mk. I and II prototypes both suffered from longitudinal instability, especially on the climb and in level flight at speeds below 190

R2268 fitted with wide span tailplane and endplate fins and rudders similar to the Bristol 153A project. (Bristol Aeroplane Co.)

---

* This was only one of many tests carried out at RAE Farnborough on the Bristol Hercules to achieve a low-drag power plant installation; the ultimate result being the Hercules 100 series power plant.

Hercules engine Beaufighter installation on test in RAE wind tunnel. (RAE)

mph IAS. To correct this the manufacturer carried out a programme, in conjunction with A&AEE, of modifications; these included a wide-span tailplane with endplate fins and rudders, similar to the Type 153A, fitted to airframe R2268, but flight tests determined that it was neither satisfactory in stability nor in performance. One further attempt to improve stability was the increase in span of the tailplane, and the fitting of horn-balanced elevators; this also failed to cure the problem. The final solution was, in the end, the 12 degree dihedral tailplane; this was initially fitted to Mk. II aircraft, but was also retrofitted to a number of Mk. I aircraft as well. Although it did improve stability, some fighter pilots and fighter squadrons frowned on this modification, as it was felt to make the aircraft too stable for fighter-type manoeuvres.

As the take-off swing on the Mk. II was initially unacceptable on a night fighter, Bristol's altered the fin and rudder shape; ultimately the solution was found with the incorporation of a dorsal fin, which gave an increase in area of 10.9 sq ft. Thus, when Coastal Command's Beaufighters had additional equipment installed, and a torpedo, and the CG moved aft of the previous acceptable limit, the same solution was necessary, and a dorsal fin introduced to improve the lateral and directional qualities of those aircraft. This was standard on the Mk. X, but was also fitted on some Mk. VI aircraft.

The Beaufighter was now in production at Filton, Old Mixon at Weston-super-Mare, Fairey at Stockport and Rootes factory at Blythe Bridge. Numerous subcontractors were manufacturing parts or sections of the aircraft, including members of the automobile industry, such as

Mk II R2058 at Hucknall fitted with Merlin XX engines and standard fin and rudder. Nascelles 'tufted' for airflow check. (Rolls-Royce Ltd.)

Briggs Bodies, Austins and Standard. Fairey Aviation, for instance, had been drawn into the Beaufighter Group in late 1939, and co-operation between them and Bristol had been excellent, as was testified in a letter from their chairman to MAP. The result of this was that Fairey's first Beaufighter, mainly comprised of Bristol-manufactured parts, was completed in November 1940, and on 11 January 1941 the DTD gave authority for the issue of Specification Beaufighter 1/P3 for the production of the Beaufighter by 'daughter' firms.

The position of Fighter Command and Coastal Command Beaufighter aircraft was more stable by the middle of 1941, and the factories producing the Beaufighter were well in their stride; so that by December the decision was made to make a complete separation of type. This also resulted in the modifications being separated into three classes; Class A modifications that were applicable to both Fighter and Coastal Command, Class B applicable to Fighter Command and Class C applicable to Coastal Command.

The first production Beaufighter to come off the production line with fixed fittings for Coastal Command was the 217th aircraft, and this was delivered during the first week of February 1941. The production of the Beaufighter required over 40,000 tools, some of which were produced by Briggs Motor Bodies. Not only was production speeded up due to the demand for the Beaufighter, but damaged aircraft were being repaired on site by the robbing of spares off aircraft at the works. A report on this would be issued in February 1942, and from this it was seen that 284 aircraft had been brought back into service this way over the past six months at an average time of 3.2 weeks per aircraft. This repairing on site standardized many repair schemes and practices; it being found, for instance, that tubular members of the nacelles, if damaged, were easier and cheaper to replace than repair. Whereas when replacing large areas of skin, it was better to commence riveting from the centre to prevent panting.

With Beaufighters operating over the sea on interception and strike, it was obviously necessary for information on the ditching of the aircraft; so in late 1942 RAE Farnborough began model tests on the alighting on the sea of the Beaufighter Mk. I and II as a background to the investigation of any actual ditching. Dynamically similar models of the Hercules- and Merlin-powered Beaufighter were constructed, and catapulted at scale flying speeds in still air onto calm water. The behaviour of the models was observed and in

Mk II R2058 fitted with 1st stage of fin/rudder modifications.

many cases the action was photographed with a high-speed camera.

From this it was determined that it was possible to ditch both marks of aircraft safely, using one-third to two-thirds flap, the deceleration being in the order of 2–2½ 'g'. The Beaufighter, especially the Mk. II, it was found, might be slightly directionally unstable in the water and the floating was expected to be short. Landing the aircraft with the wheels up was found best, as ditching with the undercarriage locked down could result in violent diving or somersaulting.

At this point there was fortunately little experience of full-scale ditching, but what there had been, supported the model test findings; although originally the opinion was that it would be a bad aircraft for ditching. One experience was that the first impact of the tail on the water was gentle, then, as the forward part of the aircraft sunk in the deceleration became very high. One report of an actual ditching referred to a Mk. II and it was reported that, 'The aircraft came to a dead stop, within 15 seconds the tail rose 50 degrees above the horizon, within 30 seconds the cockpit and wings were awash as the aircraft began to sink nose first. The aircraft sank within 3 minutes'.

With the successes during 1941 with AI, further squadrons were converted to Beaufighters, and likewise, radar development was speeded up. For whereas the AI Mk. I had been 'handmade', further developments had taken it into production status, so that by the end of 1941 the AI Mk. IV was in general use. All these early marks naturally had their failings, but such was the speed of development that by this date there were the Mks. VII and VIII under development. At the end of 1941 and the start of 1942 there were up to eight Beaufighters at TRE Hurn (including R2373, R2347, X7624 and X7712) involved in AI development. Amongst this development was testing of AI Mks. VII and VIII, operational technique for intercepting AI jamming aircraft, tests of IFF Mk. III, as well as testing of windscreen projection (H.U. Display) of both AI and artificial horizon.

The introduction of Mk. IV onto the Beaufighter was a quantum leap in performance, for although it still worked on the 1.5 metre band, a new transmitter output valve was introduced which reduced the minimum range to 400 ft, whilst the range was 4 miles at an altitude of 21,000 ft. With the introduction of the Mk. VIII, AI radar had moved into the S band (10-cm wavelength); in this set, instead of employing wave guides to carry output from the modulator to a rotating aerial, the Mk. VIII had a fixed aerial and a rotating reflector.

AI VIII had one failing, which was not compatible with the Beaufighter, for, as flight tests and operations had proved, it was not a highly manoeuvrable aircraft. The AI failing, or difficulty, was that when the Beaufighter had closed the range to approximately one mile, if the target aircraft then received warning of a Beau' up the back end, it could by means of a violent manoeuvre, escape from the sector being scanned. This not only 'threw' the RO and his AI, but the Beaufighter as well. To overcome the difficulty with the AI, there was now introduced the Mk. IX, which was a 200 kW S band radar and featured 'auto-hold'. In this model the aerial dish was motivated by electro-hydraulic servos operated by the RO, and could be pointed at the target and automatic tracking switched on; the target being tracked by feeding the azimuth and elevation signals to the scanner drive — the date was still 1942, and Beaufighters were still the main hunters of the *Luftwaffe* night bombers.

On 25 July the Air Ministry notified the DGAP that on top of their previous order for Beau's fitted with Mk. VII AI radar, they now wanted a further twenty, to cover wastage of aircraft with this equipment. Yet fear of the Mk. VII equipment falling into enemy hands, resulted in the Air Ministry later ordering MAP not to

Mk VIF X7542 with AI Mk IV radar during Boscombe Down trials.

dispatch overseas Beaufighter VI aircraft fitted with AI Mk. VII or VIII. This was later amended to Mk. VIII before May 1943 at the earliest, and the reason given was that there would be insufficient aircraft fitted with this equipment to satisfy overseas commitments.

On the subject of night fighting, the USAAF did not at this period have an effective operational night-fighter, and was relying on modified DB7/Bostons. This resulted in the U.S. Naval Attaché, Admiral Kirk, visiting the VCAS on 24 December 1942, and during their conversation he mentioned the fact that Washington might in the near future request from the British a minimum of twelve night-fighter Beaufighters for the defence of Guadalcanal, as well as giving training to both the air and groundcrews. Rumblings must have been heard at the back door(!), for not only could the RAF not get enough Beaufighters for overseas squadrons, but the replacement of the Beaufighter with the Mosquito on night-fighter duties was already planned.

Digressing a little and returning to the initial decision to fit AI radar, the experts at Bawdsey establishment* considered that the decision to install the radar operationally was premature and would lead to a lot of problems. The justification for this decision to get the AI radar into service

* Originally the radar development centre.

X7579 experimentally fitted with AI.Mk VIII radar. Old Mixton factory-built aircraft.

was the nil result of the night-fighters during the *Luftwaffe* air raid on Coventry. Initially the crews had no idea of the radar equipment or the operating technique for night interception — their experience being based on day fighting, as opposed to the stealthy stalk of the target at night. As crews gained experience with the AI Mk. IV and night interception, so the scores increased, as witness the results in Chapter 5. At the same time, better and more serviceable AI radar was introduced, with even better results.

The ultimate radar in service during the 1939–45 war was the AI Mk. X, which was a modified American SCR-720-B, which provided facilities for interception and ranging of target aircraft, for beacon and homing services, as well as use as a navigational aid. It operated on a wave-length of 9.1 cm and produced a narrow beam of radiation approximately 10 degrees wide. As far as is known only one RCAF Beaufighter was flown with a SCR-720 fitted.

It will thus be seen that airborne radar and the Beaufighter developed along parallel lines, complementing each other, both with AI radar and ASV radar. A combination of the two was tried in 1943, when a Mk. VI V8564, fitted with

AI Mk. VIII radar, was test-flown to consider the use of this radar equipment as ASV for Coastal Command Beaufighters. The results of the tests proved that the AI Mk. VIII when compared to the ASV Mk. II appeared better, and did not compare unfavourably with centimetre ASV; great accuracy being obtained at the ranges at which a torpedo was launched. Along with the recommendation for its use on Beaufighter torpedo-bombers, was the recommendation to incorporate a modified mirror.

To conclude this part of the development story, and continue on the theme of radar, it may be of interest to the reader to learn that one very senior RAF officer, Air Marshal Sir Hugh Dowding, having introduced radar into the defence scene and guided Fighter Command through the Battle of Britain, was in 1941 writing with an eye to the future, for in his communication, he stated that he saw the time when *every* fighter aircraft would be fitted with AI radar — history has vindicated this great man's statement.

Increased armament

In late 1941 the Frise design team at Bristol proposed to the Air Staff a version of the Beaufighter as a successor to the Beaufort as a torpedo-bomber; a response was made to this proposal on 13 April 1942, when official approval was given for the firm to proceed on a trial installation, capable of carrying either a British 18-inch or an American 22.5-inch torpedo. The TI aircraft chosen was X8065, which was delivered from the Weston-super-Mare factory on 17 April for conversion at Bristol. This was finalized and the factory flight trials completed by 8 May, when the aircraft was flown to TDU Gosport. Flight trials carrying the 18-inch torpedo with 60-inch Monoplane Air Tail (MAT) and gyro-controlled ailerons were carried out by Wing Commander Shaw;* these trials were covered by report Gos/Torps/6/1942, and cleared the Beaufighter VI as suitable for dropping the 18-inch torpedo at a ground speed

on release of 150–200 knots, at a height of 100–250 ft. The handling of the aircraft was considered light and responsive as loaded, and was subjected to dives of 240, 260, 280 and 300 knots, with steep turns and pull-outs from dives.

With the acceptance of the Beaufighter as a torpedo-fighter, it was decided to provide a torpedo capable of exploiting the performance and flexibility of this aircraft. To achieve this it would be necessary to strengthen the torpedo to accept higher entry speeds and to provide an improved method of controlling its flight in the air. After trials with six torpedoes which were tested to destruction, a new torpedo was produced, the Mk. XV, which could withstand an entry speed into the water of 300 knots. To improve the torpedo handling in flight the ATDU developed the Toraplane (known as Tora), which was a standard torpedo equipped with wings and

---

* Killed later when he crashed after an engine failure.

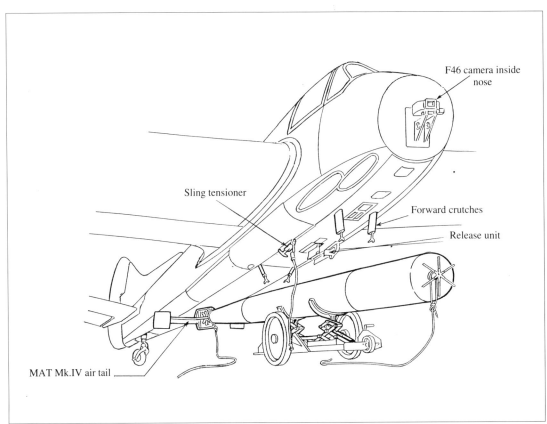

F46 camera inside nose

Sling tensioner

Forward crutches

Release unit

MAT Mk.IV air tail

Figure 2. MAT-tailed torpedo and Beaufighter Mk. TF 10, illustrating torpedo-carrying gear.

tail. After its release from the aircraft, it glided down at a speed of 150 knots, and as it entered the water the wings and tail were jettisoned.

This development was cancelled and a new air tail, MAT Mk. IV, was introduced. This was based on aerodynamic experience with Tora, which introduced the gyro-controlled air tail; this allowed release of the torpedo at speeds up to 300 knots and a height range of 60–1000 ft. Under operational conditions the drops were usually made by the Tor-Beau' at about 250 knots and 250 ft. These technical advances almost coincided with the enemy increasing the number of their escort vessels, as well as the increase in flak guns per vessel; so necessitating an improved flak suppression tactic (*see* Chapter 5).

With the improvement in performance of the Beaufighter and torpedo, coupled with the new tactics, there was later developed the Torpedo

Installation of Vickers 'S' 40 mm gun installed in starboard gunbay of R2055. (B.Ae Bristol)

Director Type F. This system employed a computer, gyro-angling power unit and the travelling lamp sight. The main function of the sight was to solve the velocity triangle and aim-off angle required from the settings fed into it, as well as to provide this aim-off angle to the line of sight. The travelling lamp sight itself on the Beaufighter was mounted at the top of the windscreen, while the mirror part of it was mounted approximately two-thirds down the windscreen, so projecting on to the windscreen the correct aim-off angle for the pilot, in the same manner as the gyro reflector gunsight.

With the introduction of the Beaufighter VI with Hercules VI engines into Coastal Command, there soon began to be complaints from the pilots that this version was not so fast at

Jettisonable rocket projectile rails on Beaufighter X NE352.

sea level, as previous versions that were powered with the Hercules XI. To counter this and to improve the performance once more, Bristol's cropped the supercharger impeller and locked the blower in MS gear; thus modified the engine was designated the Mk. XVII, which had an output of 1735 hp at 500 ft. The Beaufighters powered by the Hercules XVIIs and embodying other modifications were designated Mk. TF. 10 with torpedo, or Mk. XIc without torpedo installation. Both these aircraft in production form were fitted with a Browning 0.303-inch machine-gun on a free mounting for the observer. Prior to this, in the Middle East, squadrons had already fitted earlier model Beaufighters with Vickers 'K' (or VGO) machine-guns, as 'scare guns' for the observer; this was given authority by the issue of modification 1170. The 'K' gun modification was taken up by Home squadrons and officially adopted on production as Modification 1162, superceding 1170.

A further, and different, installation was the fitment of two 40 mm cannon in the gun bays of the Beaufighter. This occurred at the Aircraft Gun Mounting Establishment at Duxford, where R2055 arrived from Filton on 24 March 1941. A Vickers 40 mm cannon was installed in the starboard compartment, all 20 mm cannon having been removed; as delivered the gun only fired single shots, but after air firing on 11 July plans were put in hand to convert it to belt feed. On 12 August the 40 mm Rolls-Royce cannon was received, and this was installed in the port position of R2055, making it the most heavily armed fighter aircraft of that period.

This work was carried out under the guidance of Squadron Leader J.G. Munro*, Officer Commanding AGME Armament Testing Squadron, who was also involved in the flying, and had been involved pre-war with the Whirlwind cannon installation as well the belt feed installation on Whirlwind L6844 at AGME in 1941. Testing of the Beaufighter with the two 40 mm cannon was carried out in October, both on the ground as well as in the air, and both were

* Permanent Commission RAF 1934. Flying Course 1934–35. Armament Course 1937–38. R&D Air Ministry 1938–40. Squadron Commander 263 Squadron 1940–41. Chief Test Pilot AGME 1941. Wing Commander Fighters TFU 1942. Various technical posts until retirement as Group Captain 1949.

found satisfactory. The Vickers 'S' 40 mm cannon was considered the best for Service use, and was put in to production for installation on the tank-busting Hurricane.

The decision in favour of the Vickers 'S' cannon was most probably biased by the fact that Rolls-Royce did not have the armament manufacturing facilities available to Vickers-Armstrong. This weapon was recoil operated, had a long recoil and a fifteen-round magazine, and was designed to fire Naval type 2-pounder shells of AP or HE; but initially only fired solid shot and single shots. This was redesigned for magazine feed and fired 125 rounds per minute; it weighed 295 lb, had a muzzle velocity of 1800 ft/second and an effective range of 2500 yards. Whereas the 20 mm shot would punch holes in a locomotive, with the 40 mm shot the boiler left the locomotive, and removed the locomotion!

The rocket projectile (RP), originally listed as the unguided projective (UP), would be the next weapon to join the Beaufighter's armoury, and would in the end become the standard anti-shipping weapon of Coastal Command. It first appeared on Beaufighter VIC EL329 in September 1942 as a trial installation of four rails under each wing carrying 3-inch RPs. Just in case the rocket blast on firing should prove a problem to the airframe, 'blast plating' was fitted under the wings above the rails. Flight tests as well as operational use soon proved this unnecessary. Flight trials were also conducted to determine the effectiveness of the various rocket projectile heads.

The rocket projectile consisted of a rocket motor containing propellant, four fins, and two saddles for mounting; the head is either 25 lb AP or 60 lb HE/SAP, with both weights having concrete practice heads. There was also a rocket flare head developed for use in illuminating the target area at night for attacks on U-boats or surface vessels. The rocket rails also came in for development; so to determine the effect that rocket rails had on the aircraft's level speed, flight trials were carried out on NE343 and NE352; this eventually led to the jettisonable type of RP rail being tested and proved for use on the Beaufighter.

Whilst not strictly coming under the increased armament heading, yet having an effect on the placing of the armament aiming, the following is worth mentioning. During the early days of the

Beaufighter night-fighting episode, the ordinary 50 mph graticule sight was used as the attacks were usually made from the rear with little deflection; then with the interception of faster targets came the introduction of the GM2 gunsight, as used on day fighters. Likewise in regard to the approach to the target aircraft or vessel, it was sometimes necessary to slow the aircraft down, due to too fast an overtaking speed, which was easy to do at night. To overcome this problem it was suggested that the lowering of the undercarriage to increase drag and reduce forward speed was possible, as the limiting speed on the Beaufighter was quite high — obviously tried by some exploring pilot.

On 1 April 1941, 600 Squadron suggested to Fighter Command HQ, that a quick-acting airbrake for use at night was essential to prevent overshooting. This was followed through by the AOC-in-C Fighter Command asking for an investigation into a brake flap for the Beaufighter, and on 30 May Bristol's agreed to the use of lowering the undercarriage as a means of reducing the speed. To satisfy the Coastal Command pilots' need for an airbrake, Bristol's in conjunction with A&AEE, developed the split trailing edge airbrake (described in Chapter 3) — but even that brought the odd complaints, which usually centred around the fact that the installation reduced the top speed by a few mph.

What is remarkable about the effectiveness of the Beaufighter as a weapon carrier and its use in ground attack and as a strike aircraft, is that numerous authorities forecast its *ineffectiveness* in these roles. In April 1942 the DOR in correspondence to the ACAS(T) stated that the Beaufighter was unsuitable for ground attack because of its general flying characteristics. Boscombe Down had reported that the Beaufighter was not manoeuvrable, and long low approaches would have to be made for accurate shooting, and the pilot's view was poor — but what is plainly obvious to anyone who has sat in a Beaufighter cockpit, let alone flown one, that this aircraft with its well-armoured cockpit and two air-cooled engines, has a magnificent view forward and to the wingtips, thus ideal for a strike or ground attack role.

Then again in February 1943, at a Coastal Command conference chaired by Slessor, whilst accepting the Beaufighter in the anti-shipping role with torpedoes, rockets and cannon, considered that it could only carry a small and ineffective bomb-load of two 250 lb bombs — was there any information sought as to the Beaufighter's bomb-carrying ability? For in early 1944 trials would take place at Boscombe Down with a Beaufighter X NT921 carrying 1500 lb of bombs, whilst the mainplanes, originally designed for the Mk. XII aircraft, were capable of carrying a 1000 lb bomb each side.

The results of the trials of the Beaufighter with two 500-lb bombs under the fuselage, and two 250 lb depth-charges under the wings, concluded with it being declared suitable for Service use. As had already been noted at ATDU Gosport during torpedo-dropping trials: 'The power available and the high rate of climb results in a good "getaway" after release of the torpedo' — the same applied to a strike or ground attack aircraft.

The ATDU Gosport in July 1942 also highlighted the need for a rear defence gun, this had been foreseen by some squadrons installing local-made mounting for an observer-operated 'K' gun. Now was introduced the Type B20 gun mounting, which carried a Browning 0.303 inch machine-gun fed by a 500-round magazine. This was installed in the observer's cockpit hood, which was hinged on the fuselage and could be

TF.X NT921 at A&AEE in 1944 modified to carry two 500 lb bombs under the fuselage and two 250 lb bombs under the wings. (RAE)

opened as an emergency exit. The mounting had limited traverse and elevation, but at least offered a measure of defensive fire; whilst the gun could only be depressed 10°, its elevation was 35°; in traverse this was 35° each side of the centre line, and an interruptor prevented damage to the aircraft structure.

Beaufighter II and V

As production of the Hercules was lagging behind the programme, the decision was made to provide an alternative engine for the Beaufighter; this resulted in a list of comparative weights and CGs for a Beaufighter fitted with the Merlin, Griffon and the Hercules (Table 3).

	Hercules	Merlin	Griffon
Normal all-up weight	18,250 lb	17,500 lb	18,600 lb
Overload weight	19,575 lb	18,800 lb	19,950 lb
C of G position, normal	49.25	51.87	48.3
C of G position, overload	49.27	52.4	49.1
C of G position on aero- dynamic mean chord	0.292	0.315	0.283

Table 3. Comparison of three different engines on Beaufighter

As the Griffon had been allocated to the Royal Navy's aircraft programme, a crash programme was initiated to re-engine the Beaufighter with the Merlin XX. Air Vice-Marshal Sir Wilfred Freeman had been informed by the Secretary of State for Air that Rolls-Royce were anxious to build fifty power-units for the Bristol Aeroplane Co., so he wrote to Hives to state that there was no objection to this provided that it did not

IIF R2270, first production Mk II, fitted AI.Mk IV. note nose-mounted dipole transmitting aerial and shrouded exhausts.

interfere with Merlin production. This was followed on 28 May 1940 with a letter from Freeman to Hives, asking whether the supply of 100 Merlin power plants (fifty aircraft sets) would interfere with the Merlin programme. Two airframes, R2058 and R2061, were delivered to Rolls-Royce Hucknall on 30 November and 28 December 1940. R2058 was no stranger to Hucknall, for it had flown from there in the previous July, powered by Merlin Xs. Now, both aircraft were to have Merlin XX engines installed. With the incorporation of the Merlin engine and its liquid cooling system on the Beaufighter, a further problem arose: to provide armour protection for the header tank. By early July 1941, Hucknall was preparing a trial installation, which comprised an armour plate over the header tank, and thick dural sheet over the front part of the engine cowling. The aircraft powered by the Merlin engine was designated the Mk. II.

The long engine power plants and torque increased the swing on take-off, and so a trial installation of a high aspect ratio fin and rudder was carried out on R2058. So little improvement was found with this that a further installation, in the form of a large dorsal fin, was tried out and considered as improving the directional stability.

Returning to 1940, it had been decided to allocate R2059 and R2061 as well as R2058 to Rolls-Royce for conversion, but then the Air Ministry struck R2059 from the programme as it was already ear-marked for other work, so R2062 was ordered to be dispatched instead. Meanwhile at Hucknall R2058's flight trials were being delayed, because its undercarriage doors were failing to stay closed in flight, whilst Rolls-Royce were finding it necessary to fit larger engine oil pipes. A fifty-hour contract for development flying had been awarded for the conversion, but this was already running out on R2058 and no flying had been carried out on R2061; so an extension was being sought. The added protection for the glycol tank had made a substantial increase to the aircraft's weight, which had caused a rethink at the Air Ministry; who, on 19 August, issued instructions that armour protection was not considered necessary as it was not so important on a twin-engined fighter!

R2062 also did not join the Hucknall circus, it was still at Filton and had been fitted with the new ailerons and elevators, as well as AI Mk. IV

Second Mk II prototype R2061, used for the evaluation of various installations. Here the nacelles resemble Exe engine installation. (Rolls-Royce Plc)

radar, and was to be used by Fighter Command on AI trials. Before the end of 1940 this aircraft was destroyed along with a few more Beaufighters, during an enemy air attack on Filton; this not only caused damage to the works but to local housing also. The result was a certain amount of absenteeism the following day, but within twenty-four hours things were more or less back to normal.

At Hucknall R2058 had by now been fitted with the modified fin and rudder mentioned previously, and R2061 was engaged on flight trials with a pressurized fuel system. This latter aircraft would subsequently also be fitted with the high aspect ratio modified fin and rudder, and an improved production air intake, air cleaner installation and exhaust flame dampers would also be fitted and tested. When R2058 was returned to Filton it was test-flown by the local test pilots, who reported the following: there was a strong tendency for the aircraft to swing to the left as soon as the throttles were opened, and this swing was barely controllable if the flaps were lowered to 20 degrees. In general flying there appeared little difference between the Merlin-

and Hercules-powered aircraft. The Merlin engine was smooth and the temperature control of the installation appeared satisfactory and simple; whilst the view over the engine was noticeably improved, it was considered that the radiators offered minor obstruction.

The turreted fighter obviously had some sponsor, for both the Beaufighter and Mosquito were contracted to have trial installations of the Boulton & Paul Type 'A' mid-upper turret. The Beaufighter with the turret installed was designated the Mk. V, and during March 1941 two Mk. IIs, R2274 and R2306, were converted on the production line to Mk. V standard. In this version, the turret was installed just behind the pilot, blanking off the forward lower exit; the two inboard cannon were also removed, as was the observer's blister, which was blanked over. Both aircraft were flown with operational squadrons for service trials, but were not considered with favour by the pilots, partly because of the lower maximum speed, and also because of the lack of the lower escape hatch. Operation of the turret was normal, but the stability of the aircraft was found to be worse than with the standard Mk. II.

The first production Mk. II (R2270) flew at Filton on 22 March 1941, and though 447 of this mark were built, all were manufactured at Filton and formed the equipment of a number of RAF squadrons, the 307 Polish Squadron as well as some FAA Requirement Units. Pilots who graduated straight to the Mk. II got used to the beast, but pilots who had flown the Hercules-powered Beaufighter first found it hard work on take-off. The Mk. II was used operationally only as a night fighter, and the official opinion was, that it was doubtful whether its performance justified the time spent on it. Its main advantage was that by fitting the Merlin engines as a replacement for the Hercules, a larger number of Beaufighters were produced; the service ceiling was also raised by about 5000 ft. Early flights at Rolls-Royce with the Mk. IIs gave high speeds, but these were not reproduced on production aircraft, and did not better the Mk. I in respect of speed. The official report also mentioned that flame damping never reached a satisfactory standard acceptable to the RAF, and the aircraft

was never satisfactory aerodynamically, pilots having to anticipate the tendency to swing on take-off. The Mk. II's power plant was designed at Hucknall, and incorporated a wedge-shaped intermediate bay to pick up with the Beaufighter nacelle, the manufacture being carried out at Morris Motors at Cowley. The Mk. II's dorsal fin was first test-flown on T3032, and increased the fin area by 13.1 sq ft.

A qualification of the handling of the Beaufighter II was given me by Group Captain John Wray DFC*, a very experienced pilot who had flown a wide variety of fighter aircraft:

---

* Commissioned in the RAF in the 1930s, and flew a wide range of aircraft; Blenheims in France, Beaufighters, Whirlwinds, Hurricanes, Typhoons, all marks of Spitfire, and some jets. Served thirty-two years in the RAF including a spell at NATO. He does not mention the fact that as a Wing Leader on Tempests he shot down two German Me.262 jets that got in his way.

Figure 3. Bristol Beaufighter Mk. X.

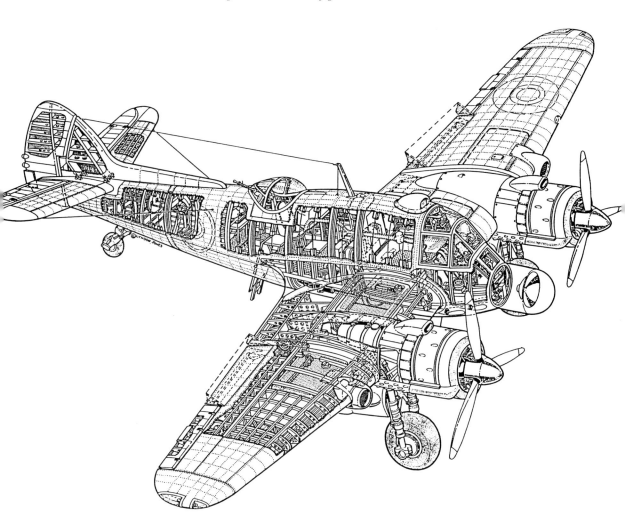

R3177 airframe converted at Hucknall to accept Griffon IIB engines. Crashed at Driffield 2 April 1943.
(Rolls-Royce Plc)

I flew four sorties on the Beau' II with Merlin XX engines; I was sent by my squadron, at that time equipped with Beau' IF., to do a comparison test. The aircraft were with a Canadian night fighter squadron at Coleby Grange, near Boston, Lincolnshire. Most noticeable was the tendency to swing violently on take-off. In the air the visibility from the cockpit was marginally better because of the lower profile of the engines. The aircraft was also slightly faster, although this was removed when we got the Beau' VI. The rate of roll on the aircraft I flew seemed to be better. At night the take-off swing had to be watched carefully. In the air there was slight flame flare from the exhausts, but one did not suffer the same problem we had on the Beaufighters with the Hercules, where throttling back quickly (often an operational necessity) caused a great deal of flame flare . . . it was not regarded as the equal of the Hercules Beaufighter, particularly when the Beau' VI arrived.

In regard to liquid-cooled engines, whether with leading-edge or nose radiators, they were susceptible to damage by debris shot off the attacked aircraft, because of the closer firing range of the fighter at night. Whilst proponents of the Merlin may claim that it could be run for long periods outside the operational limits, it could as equally be claimed that the Hercules, or any air-cooled radial engine, could sustain more damage and keep operating. These and the other factors mitigated against the Beaufighter II as a success in night fighting; it also needed to be flown at high boost and rpm to make it perform.

Whilst R2058 went on to serve with 25 Squadron, eventually to retire as instructional airframe 3344M, another Mk. II, T3032, was used for the initial flight-testing of the long dorsal fin. Another Mk. II, T3177, was selected and went to Rolls-Royce Hucknall as a flight test-bed for the Griffon IIB, but the testing of this combination was soon terminated, as the performance of the Mosquito with Merlins was more impressive — and at a greater altitude.

Other projects

The Bristol Design Office, during the period 1939–40, was considering further development of the same theme, one being a three-seat bomber version powered by the Hercules engines and mounting a dorsal turret; this was designated the Type 157. There was also an improved fighter version — known within the firm as the 'sports model' — that had a slimmer fuselage, with the cockpit raised above the decking (as in the Brigand in later years); this was the Type

158. Neither of these versions went into production when Beaverbrook became MAP 'War Lord', but in the emergency aircraft programme, the Beaufighter was given lower priority than the five aircraft types listed.

The Beaufighter Mk. VII fitted with Hercules 26 engines and the Beaufighter Mk. VIII and Mk. IX fitted with Wright Cyclone GR2600 engines, were allocated for Australian production. At least two Beaufighters were converted in Australia to Wright Cyclone power. Australia was interested in the Beaufighter from 1939, and on 1 August 1939 made known their decision to purchase eighteen aircraft, but these would require a number of alterations; they included a P4 or P6 compass in the observer's station, as well as a navigation table or equivalent, different radio sets and also provision for stowage and launching of sea markers/flame floats/smoke floats. By the middle of November the Beaufighter mock-up had been altered to suit the Australian requirements. The programmed delivery times were scheduled for eight aircraft by April 1940, and the remaining ten by July 1940; the first batch were 26 to 33 machines inclusive from the Bristol contract 983771/39. The total cost of the Australian order was £198,000 at £11,000 per aircraft. With the failure of the Beaufighter to live up to Bristol's

performance figures, and the problems encountered on production, on 18 March 1940 the Ministry was considering postponing the delivery until operational experience had been gained. Then on 1 May the Australian Air Liaison officer requested the Air Ministry to accept cancellation of the Australian order, which at that time, probably saved a few red faces at the Ministry.

Later on in the year Australia made the decision to manufacture its own equipment following the fall of France in June 1940, when the British Air Ministry made it known that it could not supply Australia's military needs while the RAF was re-equipping. Under the Australian Department of Aircraft Production, the Beaufort Division implemented a programme to produce two Bristol aircraft, the Bristol Beaufort G.R and the Beaufighter. With engines unobtainable from England, it was decided to set up an engine factory in New South Wales, the engine to be produced being the Pratt & Whitney 1200 hp Twin Wasp. In spite of what had been previously stated by the British Air Ministry, parts and certain equipment were still sent out from Great Britain. The Bristol Aeroplane Company also

Australian modified Fairey-built A19-2 fitted with Wright Cyclone R2600 engines. (Keith Meggs)

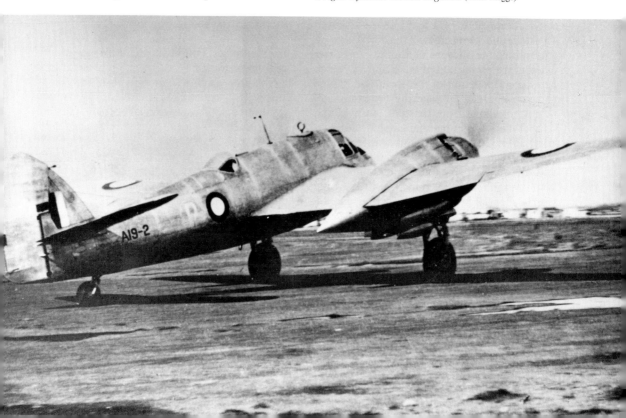

Production line of Beaufighters at Fishermens Bend, Victoria, first Australian Lincoln in background.
(Keith Meggs)

sent out the nucleus of a team, that combined with Australian engineers, thrashed out an organization; the success of this organization resulted in the production of the Beaufort, which led to the introduction of the Beaufighter. The first batch of Beau engineering drawings and data did not arrive from Bristol's until March 1943, but the Divison put their first Beaufighter into the air on 26 May 1944; this was test-flown by Flight Lieutenant J. Miles, who was OC Test Flight of No. 1 Depot, Laverton. Components and sub-assemblies were routed in to the final assembly shops at Mascot and Fishermens Bend, from where the completed aircraft were dispatched. The first Beaufighter was delivered to the RAAF on 31 May and quantity production began in September of the same year, with the Hercules power plants being sent direct from England.

This type of Beaufighter was designated the Mk. XXI, of which 365 were built, the last one (A8-365) being taken on RAAF charge on 10 January 1946. The type was initially required to be built as a torpedo-bomber, but the torpedo rule was abandoned early on. The production was not a straightforward build of the Beaufighter, for major modifications were embodied to make it suitable for the RAAF use and the tropics. Four 0.5 inch machine-guns were substituted for the six 0.303 in. ones, a rearward-facing free machine-gun was installed for the observer, pneumatic cocking of the cannon was substituted in place of manual cocking, tailwheel fitted with anti-shimmying device, provision made for Sperry autopilot, and the electrical and radio installations completely redesigned. Just in case a shortage of engines occurred, a trial installation of Wright Cyclone engines was made on two ex-RAF machines.

In Britain during 1941 a further project was under consideration at the Filton design office; this was a Beaufighter converted to a bomber to carry a 4000 lb HC bomb. It was established that the design of the fuselage around the bomb-cell would be completely new and that special bomb-doors would have to be fitted, but no standard Beaufighter parts would be discarded. In the end, of course, this did not materialize as such, but was further developed into the Buckingham, which like the Beaufighter was superseded by the Mosquito. The Beaufighter theme was carried further with their project to Specification B7/40, the two-seat light bomber Type 161; later on the Type 164 to Specification H7/42 was proposed, which was developed into the Brigand strike/light bomber, which superseded both the Beaufighter and Mosquito in the Far East.

A further extension of the Beaufighter was the final wartime mark, the Mk. XII. This was intended to have a much strengthened wing and to be powered by Hercules 27 engines, and to be capable of carrying a 1000 lb bomb under each wing. The Hercules 27 was fitted with the Bendix pressure carburettor as the Hobson was in short supply, and though large orders were placed for the Bendix type the project was cancelled. The Mk. XII wing still went into

production and was used on the later series of the Mk. X aircraft, whilst the Hercules 27s were set aside. Although the Bendix carburettor had no boost control, it was considered that this was satisfactory as the TF XII would not operate at high level — in the end it did not operate at all.

A further Beaufighter development had been projected in 1940, this was a high altitude version powered by the turbo-blown Hercules VIII, but as the aileron response above 30,000 ft was marginal it was obviously a non-starter, more so when the Hercules VIII engines proved to be unreliable. The high altitude fighter was eventually solved by the development of the Spitfire and Mosquito; as well as the issue of

'Red Cliffs', one of a late production Beaufighters paid for by subscriptions and allotted town names. Shown at Laverton. (Keith Meggs)

45

Specification F7/41, which bred the Westland Welkin.

Projects from the Bristol design office based on the Beaufighter, or its structure, continued to flow; but these gradually grew heavier, or the specification was changed, so that the theme changed from the strike fighter to the medium bomber. In the end the only hardware from these projects were the Buckingham, Buckmaster and Brigand. Then with the Air Ministry's decision to standardize on the Mosquito, there was no further use for the Beaufighter or development. Yet it would be the Beaufighter that 'wore well' under all conditions of extreme climate — not for nothing was she called the 'Mighty Beau'.

A8-1 first production Australian Beaufighter produced at Fishermens Bend. Here at RAAF Laverton.
(Keith Meggs)

Chapter 3
Testing and Analysis

Testing at the A&AEE commenced on the third prototype, R2054, in September 1940. This aircraft had the original type of nacelle in which the main wheels still protruded when retracted, and it was powered with Hercules II engines. Further development continued and a modified windscreen was fitted; in this condition and with the tailwheel locked down R2054 went for test in February 1941. The Aeroplane and Armament Experimental Establishment raised test reports A&AEE/758 to cover the tests on the Beaufighter, the 3rd Part of which covered speed measurements on R2054, R2060, T4623 and X7540 at the A&AEE, and X7542 by the Bristol Aeroplane Company. All recorded widely different speeds, varying from 300 to 335 mph (Table 4).

Serial number	Top speed	Full throttle height, ft	Remarks
R2054	309 mph	15,000	Built by Bristol. Hercules III. Original nacelle, wheels protrude 3–4 in. Very poor finish. Retractable tailwheel.
R2054	300 mph	14,600	Modified windscreen. Camera gun fitted and tailwheel locked down.
R2060	323 mph	14,400	Built at Bristol. Hercules II Flush riveted, all possible leaks sealed.
T4623	335 mph	15,800	Built at Austin. Hercules X. Redesigned air intake. Flush riveted, undercarriage doors fully close.
X7540	333 mph	15,500	Built at Weston-super-Mare. Hercules X. Redesigned air intake, Flush riveted, under-carriage doors fully close.
X7542	321 mph	15,500	Built at Weston-super-Mare. Hercules X. Redesigned air intake, flush riveted, undercarriage doors fully close.

Table 4. Speed measurement comparison

The 5th Part of A&AEE/758 covered a brief performance, handling and diving trial on two Shadow Factory-built aircraft compared against a Bristol-built aircraft. These indicated that the handling qualities were similar over the three, even though the rates of climb, top speed and service ceilings did vary slightly.

When R2270 went for test at Boscombe Down in August 1941, it was loaded for the test with the CG at 49.1 in. aft of datum. In this condition the aircraft was unstable on the climb and in level flight at speeds below 190 mph IAS; although in level flight at higher speeds the stability was satisfactory. It was considered at this stage that further tests should be carried out with a tailplane having dihedral. These tests took place in September with R2270 fitted with a new $12^{1}/_{2}°$ dihedral tailplane, similar to the one that had been fitted and tested on R2057. This was introduced because of the longitudinal stability of the night fighter Beaufighter being criticized as inadequate, so Bristol's had introduced the dihedral tailplane in an effort to take the ends of the tailplane outside the concentrated wake of the propellers. The tests showed that the aircraft as tested was longitudinally stable down to a speed of 140 mph IAS; on the climb however the aircraft was stable at its best climbing speed of 140 mph IAS, but difficult to trim accurately. Though there was slight instability on the glide with flaps and undercarriage 'up' at 120 mph, with the flaps and undercarriage 'down' the aircraft was stable at its best approach speed of 100 mph IAS; never the less, it was recommended in the 7th Part of A&AEE/758 that these tailplane should be retrospectively fitted, but that further improvements in stability should be proceeded with for the night fighters.

Diving trials were carried out on R2052 and R2054 as part of the 6th Part of the report. These were carried out at a maximum loaded weight of

*RAE designed ejector exhaust for Merlin XX on
Beaufighter R2391.* (RAE)

18,940 lb, up to a diving speed of 400 mph IAS.
On the prototype, R2052, 'hunting' was
experienced on the elevators at speeds above 230
mph IAS and diving from 16,000–17,000 ft; this
caused fore-and-aft pitching, which was more
pronounced at 340–360 mph IAS. R2054 did not
experience any 'hunting' at all.

Single-engined performance with the
Beaufighter improved over the years, but even
when installed with the earlier DH Hydromatic
non-feathering or DH 20° constant-speed
propellers, the power of one engine was still
adequate for a safe return to base. The first trial
with these types of propellers was carried out in
September 1941 on R2063. The same aircraft
was used in February 1942 when DH
Hydromatic propellers with CSUs were
the subject of the 12th Part test report; the
tests showed that in comparison with the
previous trials, there was a much improved
performance.

A Mk. I (R2268) powered with Hercules X
engines was the subject of the 9th Part of
A&AEE/758; the object being to determine the
improvement, if any, in longitudinal stability of
the Beaufighter fitted with an unshielded horn-

balanced elevator. It was the opinion of the pilots
who tested the aircraft, that the phugoids
resulting when the control column was released
were steeper than usual, and that the aircraft was
more critical to the tail trim; whilst the
improvement to the stability was slight. With the
improvement to the Mk. I and II aircraft when
fitted with the dihedral tailplane, it was not
considered worthwhile continuing with the horn-
balanced elevator.

Flame-damping trials were carried out on both
the Hercules and Merlin engine installations;
although the Hercules with the standard Bristol
pattern of a tapered, shrouded, slotted type was
considered satisfactory from a visibility aspect,
and conformed to the requirements; in action,
when 'pulling off' power quickly there was
flame flare. The Merlin flame-damping trials
were covered in a number of A&AEE reports,
and included RAE Farnborough as well as Rolls-
Royce designs. The RAE design was an air-
cooled ejector pipe with fish-tails over the wing
leading edge; this was not satisfactory and the
Merlin-powered Beaufighter went into service
with a shrouded type. R2270 was the subject of
one of these trials, with the final type tested
being a saxophone type with fish-tail exit (as on
the Halifax), and it was recommended that a
shroud should be fitted to minimize the heat

Beaufighter Mk V prototype R2274. Merlin XX engines and Boulton & Paul Type 'A' turret.

effect on the wing structure, as well as to improve the flame damping.

A trial was carried out on Beaufighter II R2274 and covered by the 1st Part of A&AEE/758a; the aircraft was fitted with a Boulton & Paul 'A' turret mounted just behind the pilot. The tests were to determine the top speed of the aircraft and the effect that the turret had on the handling qualities. The top speed with the aircraft loaded to 18,695 lb for the test was 302 mph at 19,300 ft. A slight directional swing was set up as the guns were moved to the beam, and the pilots considered that the stability was worse than with the standard Mk. II. What was not appreciated, was that with the installation of the turret the pilot's lower emergency exit was deleted, this meant that the only exit from the pilot's seat was through the roof panel or right side panel, and both were considered difficult and unsatisfactory. No mention was made of the air-gunner's exit!

For some time prior to 1942 the pilots of night-fighting and torpedo-carrying aircraft had requested the incorporation of an effective airbrake, which would be capable of materially reducing their forward speed and of producing a greater angle of dive for a given speed, without any change of trim about the three axes. The Bristol Aeroplane Co., to meet the Air Ministry requirements, carried out a series of experiments to determine the best type of airbrake; this was finally resolved by an installation development of split trailing-edge type airbrakes. The aircraft chosen for this installation was R2057, which was the second production Mk. 1F from the Filton production line. After the standard production aircraft test, R2057 went to the A&AEE for trials and further development of the airbrakes, which took part in three phases.

The first phase covered trials with the brake flaps positioned on the underside of the wings and forward of the landing flaps, but after the flight test the firm's test pilot complained of large changes in longitudinal trim as well as tail buffeting when the flaps were lowered. With the second phase the brake flaps were moved to a position with one airbrake flap positioned above the landing flap and on the top surface of the wing, with the lower brake flap positioned on the landing flap. During tests this combination gave a slight tail-heaviness throughout the speed range, but this was not disconcerting and could be easily held by the pilot. The trouble was that if the airbrakes were open as the aircraft decelerated, a lateral rolling tendency was introduced.

The third and final phase involved a system that had been tested on a Beaufort; this was based on the Fairey-Youngman bellows type divebrake, which was of the split trailing-edge type and was held shut by suction from a venturi positioned beneath the wings. Upon selection of flaps 'open', the venturi suction was switched off and ram pressure applied. The brake flaps when tested opened quickly and evenly, and although when tested at a speed of 280 mph ASI, there was a slight change of longitudinal trim, this was easily overcome by the pilot without conscious effort. These brakes consisted of rectangular flaps 6 ft 6 in by 1 ft 6½ in. in size mounted on the top and bottom surfaces of each wing at the

R2057 dive brake installation showing lower section lowered and venturi installation. (B.Ae Bristol)

trailing edge, with the bottom brake flap attached to the lower surfaces of the landing flap of the outer mainplane. The operation of the flaps was controlled by a venturi beneath each wing. This final system was cleared by A&AEE as satisfactory for the Beaufighter. The extra drag due to extending the flaps was calculated as 350 lb at 100 ft/sec, whilst the drag of the installation on the aircraft contributed to a drop of 9 mph at the all-out speed.

Tests were made on the Beaufighter VI prototype to check that the oil inlet temperatures were within the requirements. So oil cooling tests with a modified oil cooler and duct by Marston were the subject of the 3rd Part of 758b, the trials aircraft being X7881. In level flight in FS gear at 17,000 ft and MS gear at 10,000 ft the oil inlet temperatures were found to be within limits.

A Beaufighter II with a dorsal fin, which extended from forward of the normal fin to just aft of the observer's dome, was the subject of a handling trial with T3032. The aircraft was powered by two Merlin XXs driving Rotol RS5/5 propellers with the aircraft loaded to a maximum of 20,400 lb. The trials determined

Showing dive brake installation fully open with top section held open for photograph. (B.Ae Bristol)

that the dorsal fin had no adverse effects on the handling qualities, whilst the advantages gained were fourfold; less tendency to swing on take-off, and when cutting one engine there was still yaw, but the rolling tendency was reduced; single-engined behaviour was noticeably better, more so with the port than the starboard; and the aircraft was in general reluctant to side-slip. This particular aircraft was fitted with convex ailerons and elevators, and the test pilots noticed that they were lighter than on a standard aircraft, and showed no sign of overbalance.

Next came tests with a modified tailplane and dorsal fin — these were carried out on a Beaufighter TF.X aircraft, NV451; the reason for the tests was that the incorporation of additional equipment in Coastal Command Beaufighters had resulted in the movement of the CG to a position aft of the previous acceptable limit. So to improve the stability a modified elevator having 40 per cent aerodynamic balance and including two anti-balance tabs had been fitted to NV451. A dorsal fin had also been fitted to improve the lateral and directional qualities of the aircraft. The tests concluded that the general

handling qualities of the test aircraft showed a marked improvement over the standard aircraft. The aft acceptable CG limit for this aircraft was now approved at 57 in at a weight of 24,120 lb, though the aircraft at this weight should be confined if possible to experienced pilots. A CG limit of 56 in without restriction was acceptable on account of improved handling.

In regard to this test on NV451 it was also stated: 'The presence of the dorsal fin has improved the lateral and directional qualities of the aircraft, though the residual rolling motion still present in normal level flight detracts from the suitability of the aircraft as a platform for RP weapons . . .' The Beaufighter X as a standard aircraft was cleared as acceptable for the carriage of rocket projectiles during trials covered in Parts 13, 17 to 20 of A&AEE/758c.

The Beaufighter was now becoming the multi-role combat aircraft (MRCA) of the 1940s as regards roles and weapon carrying. Cannon, rocket projectiles, torpedoes and bombs, were all

TF.X NV451 on test at A&AEE in 1944 with dorsal fin, dihedral tailplane, modified elevators and carrying eight 60 lb RPs. (RAE)

part of its arsenal; while it operated as a night fighter, escort fighter, strike fighter, torpedo fighter and general purpose attack aircraft. That deterioration in performance did not quickly occur was confirmed by handling and single engine performance tests, which were carried out on a new and a six-month-old Beaufighter VI. The new aircraft was straight from the production line, and the Service one, T5103, had been in use for six months and had flown 260 hours up until the tests. For the tests both aircraft were flown at a maximum weight at take-off of 21,415 lb. Very little difference was found between the two aircraft, and those that were, were no greater than would be anticipated between two normally identical aircraft.

A speed at height comparison between a Mk. VI and a Mk. X aircraft, with both carrying torpedoes was carried out. The Mk. VI at an all-up weight of 24,300 lb and the Mk. X at an all-up weight of 25,500 lb, indicated the speed of a Mk. VI at 6000 feet as 304 mph and the Mk. X as 299 mph. At 12,000 feet both aircraft had a similar maximum speed.

Although Hercules-powered Beaufighters were operating satisfactorily in both Fighter Command and Coastal Command, and engine handling by the pilots was 'per the book', obviously in combat continual monitoring of engines was not possible; but at altitude

there was a certain amount of rpm 'hunting' and oil being thrown out of the engine oil breathers, which apparently upset some people. Amongst the latter were the test staff at A&AEE Boscombe Down, who in carrying out their test flying, were testing the Beaufighter to its limit in all aspects of the aircraft's performance. Against this must be set the fact that the Beaufighter's altitude limit in the UK was about 30,000 feet, where the ailerons became ineffective, and combats at night were taking place well below this height; whilst Coastal Command were more interested in maximum speed low down.

With regard to the Hercules engines on the Mk. VI, there was still trouble with the oil system of the Beaufighter aircraft during testing, in that the original system proved unsatisfactory at altitude — violent propeller surging and oil spillage from the engine breathers occurred. So a redesign of the installation took place, which included the fitting of an offset-circulating hotpot. As an interim measure the oil tank hotpot was removed at Boscombe Down, and during the end of 1941 and the start of 1942 a test on this type of installation was made on X7882. This indicated that the original fault with the original installation no longer occurred below 30,000 ft, so Boscombe Down suggested that as a safety measure, the operating height should be limited to 25,000 ft until the redesigned installation was approved and installed.

Further oil cooling and cylinder head temperature tests were required on the

Beaufighter VI, so X7542 fitted with Hercules VI engines was allocated for the trials. During these trials the Hercules VI engines were re-rated to give the following power:

	Old limitations	New limitations
Take-off and full throttle level flight	2800 rpm +7 psi	2900 rpm +8 psi
Maximum climb & continuous rich mixture cruise	2400 rpm +5psi	2400 rpm +6 psi
Maximum weak mixture cruise	2400 rpm +1 psi	2400 rpm +2 psi

Table 5. Hercules VI power changes

With the new power ratings, the tests indicated that both the oil inlet temperatures and cylinder head temperatures on the climb were within the requirements, providing that the aircraft was climbed at the initial speed of 160 mph ASI with cooling gills fully open, and the oil cooler exit gaps at the stated positions. During oil cooling tests in level flight in rich mixture, 'coring' was experienced on the port oil cooler, which had the greater exit gap. It was therefore recommended that it would be necessary to close the port oil cooler exit flap for operating in the winter in the UK.

With the re-rating of the engines, as would be expected, it was found during the test (6th Part of A&AEE/758b) that the rate of climb had been increased below full throttle height, both in MS and FS gear, due to the increase in boost. In level

flight in MS gear there was an increase of $2^1/_2$ mph due to the increased rpm and boost, but in FS gear there was a decrease of $1^1/_2$ mph above full throttle height at the new limits — this latter result was due to the decrease in propeller efficiency at the higher rpm and thus a net decrease in thrust horsepower.

X7542 was also the test vehicle when the A&AEE carried out trials to determine the Beaufighter VI's maximum still air range. The aircraft was powered by Hercules VI engines, and the tests indicated that the maximum still air range was, at a take-off weight of 21,000 lb and 550 gallons of fuel, approximately 1500 miles, and at a take-off weight of 21,800 lb and 679 gallons of fuel, about 1850 miles.

During September 1942 Beaufighter VI EL223 was involved on a brief handling trial with a torpedo fitted; with this installation the all-up weight was increased to 24,300 lb. This had no detrimental effect on the handling and diving characteristics, provided that the dihedral tailplane was fitted, except for a tendency to wallowing in stick-free level flight. No appreciable change of trim was observed when releasing the torpedo at 170 knots, and the maximum true speed fully loaded and with the

Mk Ic EL223/G with first installation with a torpedo, Hercules XI engines.

Mk X TF on trials at A&AEE with 200 gallon overload fuel tank on torpedo crutches.

torpedo was 311 mph at 8000 ft with power of 2900 rpm and 48 psi boost.

Tests to determine the maximum weight at which height could be maintained on one engine was the subject of the 22nd part of A&AEE/758b, JL876 was the subject aircraft and was powered by Hercules VI engines. The tests were made with the starboard engine working, to give 'worse engine' performance. The engine conditions were 2400 rpm and +6 psi boost in MS gear and cooling gills fully open. The maximum weight at which height could be maintained using only the starboard engine on the Beaufighter VI, under standard atmospheric conditions, was about 23,000 lb. Though it must be pointed out that this was not necessarily true of another aircraft, as small variations in power could have a large effect on maximum weight.

As a number of accidents had occurred with Beaufighter aircraft, and though the above test had shown that height could be maintained on one engine, tests were initiated to determine the handling characteristics on throttling either engine. This was the subject of the 25th Part, and the aircraft was Beaufighter VI EL292. The tests indicated that the aircraft behaved much better than anticipated in all conditions of flight, and no violent characteristics were found. The report did conclude that in Coastal Command, with the aircraft flying low, a sudden engine failure could result in a yaw and slight loss of height, which at low altitude could result in the aircraft hitting the sea.

Coastal Command and torpedo-carrying was the subject of the 4th Part of A&AEE/758c dated 14 April 1943. The aircraft was a TF.X at an all-up weight of 24,000 lb, carrying a Mk. 12 18-inch torpedo. This was a brief handling trial, and the aircraft was powered with Hercules XVI engines. The maximum true air speed was 308 mph at 9200 ft in FS gear, maximum true economical cruising speed of 277 mph at 16,000 ft in FS gear. Provision was made on the Beaufighter X for the carriage of a 200-gallon auxiliary fuel tank, which was carried under the fuselage in lieu of the torpedo. The 10th Part of A&AEE/758c covered a brief handling and drop test of this installation, and the test aircraft was NE343. The take-off and climb-away were normal, but it was found impossible to trim the aircraft longitudinally for flight 'stick free'. Four dives were made, all with forward pressure on the stick, and one with a full throttle dive to 350 mph ASI; at this speed it was necessary to hold the aircraft into the dive with both hands. On the tank drop test a photographic aircraft flew as chase aircraft, and the test and photographic record showed the tank dropping nose first and moving well away from the aircraft.

Further engine cooling tests were called for on the TF.X, so for these the test aircraft was EL290 fitted with Hercules XVIIs, and at an all-up weight of 24,000 lb. On the first test made at 1000 ft, it was found that the mean temperature of the four hottest cylinders had exceeded the normal limitation. The second range of tests was made in level flight at 9000 ft, and it was found that at maximum weak-mixture cruise the temperatures of the hottest cylinders exceeded the limitations; although in rich-mixture cruise they were within the limitations, and the same thing applied to the oil inlet temperatures. The third range of tests covered engine cooling on the climb, and these indicated that the mean temperature of the four hottest cylinders exceeded the limitations, as did the oil inlet temperature; though the latter temperature could be controlled by opening the oil cooler exit louvres. The fourth range of tests was carried out in March 1943, and nose rings were fitted and the oil cooler exit louvres opened up from $1\frac{5}{8}$ in to $2\frac{1}{8}$ in. This resulted in the oil inlet temperatures being within limits under all conditions, and the cylinder head temperatures within limits on the climb and in rich-mixture level flight, but just exceeded by 4°C in the weak-mixture cruise.

Rocket projectiles became the strike weapon in Coastal Command as well as 2nd TAF, and on the Beaufighter four rocket projectiles were carried below each wing just outboard of each engine. During November and December 1944 the effect of RP rails on level speeds was covered by the 13th Part of A&AEE/758c. The test vehicle was a Mk. X powered by Hercules XVII engines. As it was proposed to incorporate a jettisonable rocket projectile installation on Beaufighter aircraft, the test was to determine whether such a complication was necessary, and was to be carried out at 5000 ft over a range of engine powers, with and without rocket projectile rails. These indicated an increase of 7 mph at 160 mph and 4 mph at 220 mph when the rails were jettisoned. Four further trials were carried out during 1944 with rocket projectiles and rails, and concluded with service clearance trials of the RP installation Mk. 6 Type I. This found the accuracy standard as satisfactory and was cleared for service, with the rails jettisoned satisfactorily in shallow dives up to 270 knots.

JL876, a Beaufighter VI, was used in May 1943 by the MAP for handling tests to determine the effect of opening the emergency escape hatches in flight. The results were satisfactory and the handling characteristics were satisfactory with both hatches open. When the front hatch was opened first there was a steady change of trim to tail heavy, but this could easily be held whilst retrimming. Slight buffeting was felt

Mk X TF on test in 1944 with Mk VI RP installation.

Mk 2 T3032 fitted with dorsal fin for handling trials. Merlin XX engines and Rotol RS/5/5 propellers.

throughout the airframe at speeds between 220–240 mph ASI, but there was no adverse effect. With the opening of the lower rear hatch, there was no noticeable effect on the longitudinal trim of the aircraft. With both hatches open there was no adverse effect on control or handling in level flight, engine on, or in gliding turns or on landing; though the cockpit was noisy and draughty.

Brief handling trials had been called for on a Beaufighter Mk. X that had been modified to carry two 500 lb GP bombs side by side under the fuselage, in addition to two 250 lb bombs or depth-charges under the wings. The modified aircraft was NT921, powered by Hercules XVII engines driving three-blade DH Hydromatic 55/15 propellers. The bomb-racks for the 500-lb bombs were supported from the underside of the fuselage between the torpedo crutches, with bracing stays. The scope of the tests was the coverage of the handling characteristics under all

normal conditions of flight, with particular attention being paid to the assessment of longitudinal and directional stability. With the racks fitted under the fuselage, entry and exit could not be made through the normal under-fuselage hatch, not even in an emergency.

At the request of the MAP the tests were made at an all-up weight of 24,370 lb. The tests determined that on the take-off there was a tendency to swing to starboard, but this could be controlled on the throttles. On the climb, as typical of Beaufighters the directional stability was poor; while level flight was not so unstable longitudinally as on the climb. At 180 mph IAS the aircraft was not uncomfortable to fly 'hands off'; although phugoid oscillations were of increasing amplitude, followed by a divergence to the dive or stall. Trimmed for 230 mph IAS, and the speed displaced by ± 10 mph, the phugoid oscillations were of decreasing amplitude. The final conclusions of the test were, that with the Beaufighter in this condition and military load, it was satisfactory for Service use; although the poor longitudinal and directional stability, particularly on the climb and glide, made it unsuitable for all but very experienced pilots to fly on night operations.

These tests quoted, are but a few of many tests and trials made at the A&AEE Boscombe Down, and are typical of those applied to all Service aircraft. Other tests were also carried out at RAE Farnborough, some of these being carried out on R2066, R2391 and EL161. R2066 was involved in icing trials during late 1941, which meant flying with water being sprayed into the engine air intake at 20,000 ft. During the test the air intake iced up at high rates of flow (approximately 10 gph), but the engine kept operating. It was pointed out that this was too extreme a condition, as under the worst conditions of flight the flow would be about 4 gph, and under these test conditions the hot air intake system operated satisfactorily. It was noted that the water spray appeared to have an undue cooling effect on the cylinders in front of the hot air intake.

Another Beaufighter involved at RAE was V8341, a Mk. 1F, that was flown on a comparison basis with a Douglas Boston III for a series of tests covered by RAE Report Aero 2003. The tests were to check the flight characteristics of static and dynamic longitudinal stability of two twin-engined monoplanes, to help in the determination of specifications and relating to the pilots' opinions and to the behaviour of the aircraft in flight.

It will be appreciated from the past text, that in the summer months of 1941, the demands for the Beaufighter were such, that up to twenty aircraft were reserved for various tests and trials. This was not only due to the need to 'shake out' the faults, but because it was also a period of great change as regards equipment, and especially radar. The Beaufighter was in great demand for various roles, and over the years, the need to include various pieces of equipment and armament caused an increase in weight and changes in the CG. There were also changes to improve handling, performance and stability; all these required a prolific amount of testing, which involved Bristol's and Rolls-Royce as well as the A&AEE and RAE Farnborough. So it is hardly surprising that so many aircraft were required for these trials, and why so many Service and civilian pilots and test observers were involved in making the 'Mighty Beau' a multi-role weapon-carrier — in fact, THE most successful weapon-carrier, as the chapter on operations will illustrate.

Chapter 4
The Beaufighter Aircraft

The fuselage

In general the fuselage is of all metal monocoque construction and built in three sections, front, rear or main fuselage and stern frame. The fuselage as designed and originally produced differed very little from the final production version, although the internal equipment was updated, changed or added to for various roles, and detail changes were made. It is constructed of mainly light alloy formers, mostly of lipped 'Z' section; light alloy beaded angle section stringers and covered by a light alloy skin. The formers are notched to clear the stringers, but are not attached to them. The longerons are fitted port and starboard, and two substantial keel members extend from the rear fuselage forward to former 58 in the front fuselage. The skin is of light alloy panels fixed to the formers and stringers by countersunk aluminium alloy rivets.

The front fuselage houses the pilot and his equipment, and has two longerons fitted each side, with a non-magnetic bullet-proof armour plate from the edge of the windscreen forward to former 58. The blast tubes for the four 20 mm cannon are mounted under the floor outboard of the keel members. The framing of the pilot's cockpit is improved considerably from that as introduced, with the thick framing reduced and a reduction in the number of transparent panels — or as one pilot described it: 'away from the leaded light to greater visibility'.

The rear fuselage houses the observer and controls and his equipment; the construction is generally similar to the front fuselage, except that three longerons are fitted each side with the keel members extending the full length of the rear fuselage. The rear defence gun (later fitted) and its mounting is designed to form an integral part of the observer's hood, and differs considerably from the original hood; the hood

can be used as an emergency exit. Armour plate is fitted on the front face of the rear spar and extended above it to the underside of the hydraulic tank in the roof. Armour plate is also fitted to the aft face of the rear spar between the booms, and an armoured bulkhead is fitted aft of the observer's station, sloping down from former 183 to approximately former 194.

The stern frame construction is similar to the rear fuselage, with, between the formers 249–258, a bridge piece at the top and a bracket at the bottom, which carry the vertical rudder and elevator countershaft. Formers 308.25–310.5, 330–332 and 298 were so constructed to provide attachments for the fitment of the tailplane and the tailwheel assembly, all of the formers being of built-up construction of light alloy. At the rear end the stern frame terminates in a stern part of built-up channel section, that carries two bearing brackets. The attachment of the three fuselage sections is by the ends of the skin plating being butt jointed together and secured by setscrews and Simmonds nuts to butt-straps underneath.

Two entrance hatches, which also serve as parachute escape hatches, are installed, one just aft of the pilot and one forward of the observer. On the TT10 version the rear hatch can only be used for the launching of the targets, and so the winch operator has to use the forward hatch. Whilst on the Mk. V with the mid-upper 'A' type turret, the pilot's only escape is out of the top hatch, as the turret blanks off the lower escape hatch — which is hardly conducive to good morale for either pilot or gunner. The two entrance hatches incorporate an entry/exit ladder, and both are openable in flight, extending downwards and so providing a modicum of wind-break upon baling out. Upon opening and lowering, the hatches stayed in the open position, having little effect upon the trim or handling characteristics of the aircraft.

Though the fuselage was designed initially for a crew of two, pilot and observer (radio navigator or radar operator), provision was made for a third crew member if modification 106 or 1086 was incorporated. An early modification to the fuselage structure was the incorporation of 10 SWG skin forward and aft of the observer, whilst numerous other modifications brought in or deleted items of equipment, or changed the tailwheel strut and fittings.

Mainplane

The construction of the mainplane did vary, both during design and its manufacture, as explained in Chapter 2, and during its operational career as various roles were found for it or offensive stores attached to it. In general however it is a conventional two-spar metal-covered structure, which tapers in chord and thickness from root to tip; it is built in three sections, centre plane and port and starboard outer planes. The centre plane spars are continuous through the fuselage and are butt jointed to it, the two spars having an Alclad web joined to top and bottom extruded light alloy booms. Alclad ribs and extruded beaded angle section stringers have a light alloy skin attached to them.

The outer planes are similar in construction to the centre plane, and are attached to it at the spars; the spars being of similar construction to the centre plane spars. One of the first modifications regarding the mainplane was an increase in the thickness of the skin, and an increased width to the wing joint fittings. Later on the wingtips were strengthened, then the mainplane top skin between ribs 1 and 6 was increased from 24 SWG to 22 SWG. Strengthened wings to allow the carriage of 1000 lb bombs outboard of the engine nacelles were introduced as standard on the Beaufighter TF.X;

Figure 4. Bristol Beaufighter Mk. II.

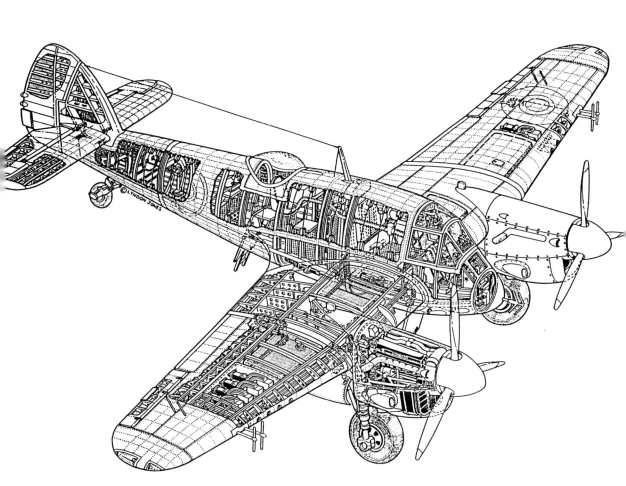

the bomb carriers were secured to the mainplane by eyebolts between ribs 5 and 6; these were introduced on to the TF.Mk. X by modifications T128–129.

The wing flaps are constructed in four sections of port and starboard centre-plane flaps and outer flaps, all hydraulically operated and interconnected by balance cables. The construction is mainly of Alclad with a 'Z' section spar at the forward face, channel section ribs with channel section transverse stringers at mid-chord, and a rectangular tube trailing edge. The ailerons are build up of light alloy ribs on a duralumin spar with a nosing and trailing edge of light alloy sheet, the rest fabric-covered. Mass balance-weights are fitted in the nosing on each side of the centre hinge. The port aileron had a fixed tab and the starboard aileron an adjustable trim tab.

A dinghy is installed in a stowage in the port centre plane aft of the rear spar.

Tail unit

This comprises a cantilever tailplane and fin, rudder and elevator. On later aircraft the tailplane has pronounced dihedral introduced, and also a dorsal fin introduced to improve directional stability. The tailplane is built in one piece with two Alclad channel section spars with extruded light alloy booms, flanged Alclad sheet ribs and skin. Modifications were introduced to the tailplane inner bearing bracket and alterations made to the size of the tailplane area as well as to the elevator. The elevator is built in two separate halves, each having a steel tubular spar, with Alclad ribs, and light alloy sheet riveted to the nose of the ribs to form a 'D' nose. The trailing edge is of oval tubular section, the whole elevator fabric-covered.

The fin is built around a front and rear post of light alloy and a rear channel section member, with horizontal ribs covered with light alloy sheeting. The front and rear posts are reinforced at their lower ends to form box sections, to which the steel-bushed fin attachment lugs are bolted.

The rudder is constructed of light alloy ribs carried on a light alloy tubular spar with a flattened extension piece at the top. The trailing edge is of mild steel tube above the trim tab and light alloy below. The whole rudder is fabric-covered and hinged at the sternpost at two points by double split-bearing hinges.

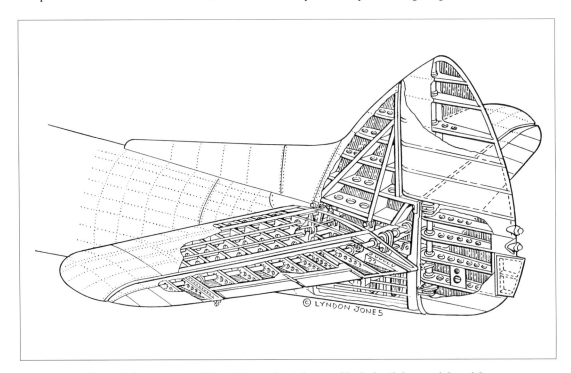

Figure 5. Construction of Beaufighter tail unit having dihedral tailplane and dorsal fin.

Undercarriage

Comprised of two independent main undercarriage units with a tailwheel unit, all are raised simultaneously by hydraulic power, the main wheels retracting backwards and upwards into the engine nacelles and the tailwheel unit retracting forwards and upwards into a recess on the underside of the stern frame.

Each main undercarriage unit has a single wheel carried on an axle mounted at the base of two shock absorbers, that are cross braced to form a rigid frame, which is connected to the engine nacelle structure at pivot points. Each unit is braced in the down position by a pair of knee-jointed radius rods, the upper halves of which are fixed to a transverse torque shaft supported in bearings on the rear tube of the nacelle structure. The lower halves of the radius rods are pivoted on brackets on the oleo legs. The shock absorbers may be of either Vickers or Lockheed manufacture, Modification 175 introducing a strengthened Lockheed undercarriage. Also introduced during service was a strengthened attachment for the oleo to nacelle structure — the all-up weight had been originally quoted as 15,550 lb, would enter service at 19,600 lb, and on the Mk. X would be increased to 25,400 lb.

Illustrating flame damper exhaust, DH Hydromatic propellers and main undercarriage oleos. (Paddy Porter)

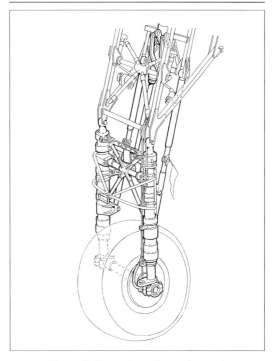

Figure 6. One main undercarriage unit.

The tailwheel unit was originally of Lockheed design, and retracted upwards and forwards into the stern frame, with the wheel forward of former 298. The oleo strut pivots on brackets on former 310.5 and has a top anchorage on the upper portion of former 298. An alternative tailwheel unit is the BLG oleo-pneumatic, which is of the trailing link type, and was introduced on the Beaufighter I and VI by Modification 42. A further introduction is tailwheel strut AIR31428 with a Marstrand twin contact tailwheel to prevent shimmy; whilst on the Mk. X a tailwheel lock for Coastal Command was introduced and a strengthened retracting mechanism at former 298.

Armament

The first fifty Beaufighter I aircraft were armed only with the four 20 mm H-S cannon; this also applied to the long-range fighter versions, where extra fuel tanks were installed in the wings in place of the 0.303-in. Brownings. The early Beaufighters were also fitted with drum-fed cannon, the drums holding 60 rounds of ammunition, and were awkward and heavy to change in combat in the confines of the cabin; the radar operator being the gun loader — amongst other duties. Only the first fifty Beaufighters were intended to have drum feed, but due to a lack of foresight on the part of MAP armament experts there would be approximately

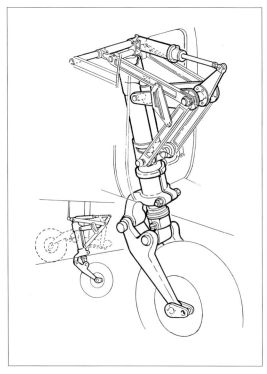

Figure 7. Tailwheel unit of BLG type.

Figure 8. Beaufighter's 20 mm Hispano-Suiza cannon layout with belt feed.

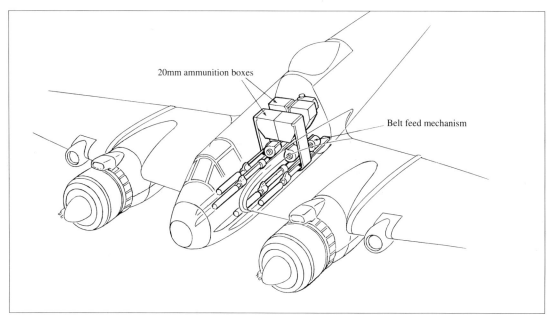

20mm ammunition boxes

Belt feed mechanism

twelve months lost before the introduction of the belt feed. This occurred because of unacceptance of a proposed recoil-operated servo feed by Bristol's. The MAP rejected the proposal on the grounds of damage to the cannon after a short period of service, or that it would jam the gun. Bristol's then developed an entirely new servo feed mechanism that used a compressed air motor, extracting the rounds by their rims. This was installed on the four cannon of Beaufighter R2053 (Chapter 2) and after inspection by the MAP armament specialists, Bristol's were informed that this would be accepted and introduced on the fifty-first aircraft, subject to it passing air-firing trials.

Meanwhile a design was brought to the UK from the Chatellerault Arsenal by two Free French officers who had escaped from France. This design was accepted as the 'Mark 1 Feed' by MAP, and went into production for the RAF; extraction of the round from the belt being made by pushing on the shell's nose. Apart from the delay in the change of official policy, there was to be delay in the production of the Chatellerault design and a change to *round extraction*. This resulted in the Mk. 1 Feed going into service on the 401st Beaufighter, which meant that 400 Beaufighters had drum-fed cannon, as opposed to the fifty originally planned for. That the rejection of the Bristol design was unnecessary was found when comparison was made with the Chatellerault design, which was found to be almost identical to the original Bristol design!

After the first fifty aircraft, all other Beaufighters had a standard armament of four 20 mm cannon and six 0.303 in. Brownings, two of the latter being mounted in the port wing and four in the starboard wing; the wing guns and ammunition boxes were reached through panels in the top surface of the mainplane. The cannon were loaded from inside the fuselage, where a gun control unit was also installed that controlled the selection of the air to the cannon. Large panels each side of the lower fuselage allowed complete access and removal of the cannon, the outer cannon being staggered aft of the inner guns.

Systems

The hydraulic system was powered by pumps on both engines, and for an emergency a hand-pump was provided. The system was so arranged that a hydraulic power lever had to be moved to 'On'

before any selector lever could select a service. Moving this power lever to 'Off' allowed the engine-driven pumps to idle over.

Hydraulic power operated the main and tail undercarriage retraction, and operation of the landing flaps. Pressure from the pumps was directed through relief and non-return valves to the selector, then on to the service, before being directed through a filter on the return line. The system operated at a pressure of 1200 psi.

The fuel system had two main fuel tanks in each wing as standard, but for long-range operating units there were two long-range tanks in each wing. These latter tanks were incorporated under Modification 605–608 action, and increased the fuel tankage to 682 gallons. Selection of the tanks was by two hand-wheels installed on the port side of the cockpit, approximately level with the pilot's elbow. Some later aircraft had selection levers in quadrants instead of the hand-wheels. The selector controls operated three-way cocks in the engine nacelles, whilst a pilot-controlled crossfeed cock was positioned in a crossfeed fuel line. Fuel jettison controls were positioned on the control panel and operated jettison valves on the two main fuel tanks in each wing, feeding to a single jettison pipe in each nacelle.

The pneumatic system was supplied from a Heywood compressor fitted on the starboard engine; the system operated the mainwheel brakes and the fuel jettison valves. The compressed air from the compressor passed through an oil/water trap, then an auto regulator to the air cylinder; from there it was cleaned in an air filter to a reducing valve, which reduced the pressure down from 450 psi to 220 psi, before passing to the services. One of which was the gun selector for the cannons, which was controlled by the AI radar operator, the air then passing to the cannon firing actuators.

A dual relay valve was provided to allow differential and progressive application of the wheel brakes; this was operated from the rudder bar, and supplied the air in the correct proportion selected to the bag-type airbrake units, two individual ones to each main wheel.

Engine

The main power unit of the Beaufighter was the Bristol Hercules air-cooled radial engine, which was a fourteen-cylinder, single-sleeve-valve

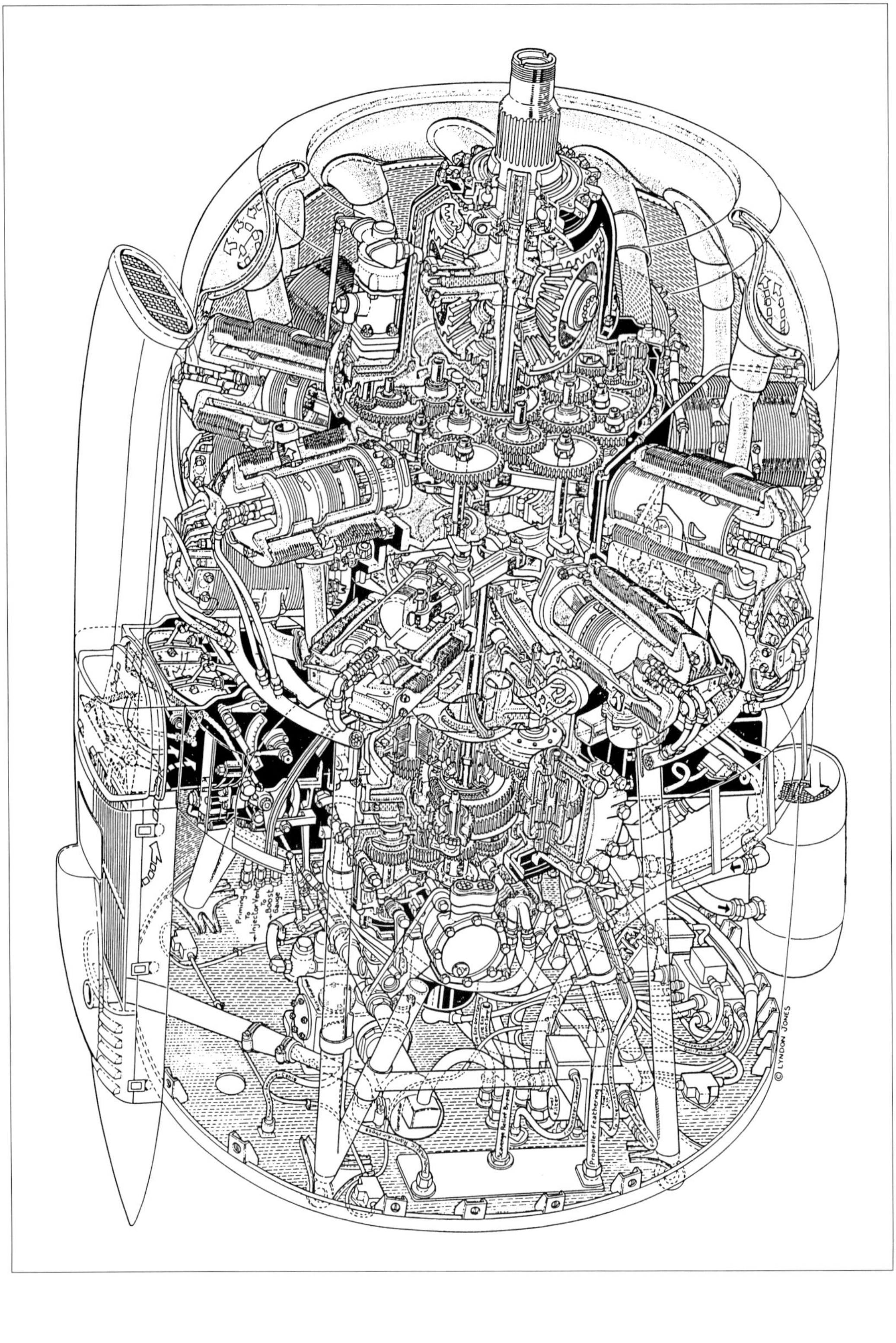

Figure 9. Bristol Hercules engine (not Beaufighter installation).

Frontal view of basic Hercules engine, which shows the cleanness of the design.

type, tightly cowled in a NACA style wrap-around cowling (basic details in Appendix 4).

The single-sleeve-valve was developed by a Scotsman named P. Burt and a Canadian named McCollum, who filed a joint patent on it. At Bristol Engines, design studies under Fedden did not commence until 1926, when Fedden foresaw the limitations that pushrod-operated overhead valves would impose on engine speeds. This was strengthened by Ricardo's sleeve-valve experiments and Barrington's study of sleeve-valve arrangements. So at Bristol's many thousands of hours of testing were carried out on a single-cylinder unit, followed by full scale testing; test units were run in competition with single-cylinder poppet-valve engines.

These tests convinced Fedden that the sleeve valve would solve problems inherent in the poppet-valve engine, he was also aware of the problems involved in developing a practical and reliable sleeve-valve engine. This would involve satisfactory sealing of the sleeves, 100 per cent roundness of the sleeves, transfer of heat from the sleeves, as well as lubrication. It took Bristol's five years to resolve the problems in producing the single-sleeve-valve engine, as well as an expenditure approaching £2,000,000.

This work involved other companies as well; Firth-Vickers produced a high-expansion steel, which could be centrifugally cast and then nitrided; High Duty Alloys perfected a special light alloy, from which could be forged cylinders with a very low co-efficient of expansion. At Bristol Engines in the special pre-production manufacturing section, Whitehead found a way to grind the sleeves truly 100 per cent round.

The first Bristol single-sleeve-valve engine was the Perseus, which had its first run in July 1932. During service further problems then manifested themselves; these related to the reduction gear planet gear retention, sleeve crank and junkhead — all of which would reappear again during the service use of the Hercules engine. Yet none was any worse than those being experienced during development of the new range of poppet-valve engines of the 1930s. Following the Perseus would come the Bristol Aquila, another single row engine, which first ran in 1934; both these were nine-cylinder units, though of different cubic capacity. Resulting from these there came the Hercules, the first Bristol two-bank engine to go into production.

The Hercules had a cubic capacity of 2360 cubic inches against the Merlin's 1649, and its single-sleeve-valve completely surrounded each piston and had a helical movement inside the cylinder bore. While the sleeve had four specially shaped ports around and near the top, the cylinder barrel had two exhaust ports at the front and three inlet ports at the rear. As a result of the reciprocating and rotary motion imparted to the sleeve by the gear-driven sleeve crank, the sleeve ports traversed the cylinder barrel ports in the correct sequence to give exhaust and inlet opening and closing cycle of operation. As there was no overhead valve gear the spark plugs could be positioned in the cylinder head (junkhead) at the optimum position at the top.

The sleeve-valve engine, in deleting the hot exhaust valve, reduced the danger of detonation, and in deleting the overhead valve gear and pushrods eliminated the need for general maintenance and setting of tappet clearances during service; at the same time reducing the total number of parts in the piston engine.

The initial testing of the Hercules commenced in January 1936, and as the HE1S produced 1290 bhp, with early flight-testing being carried out using a Northrop 2L monoplane registered

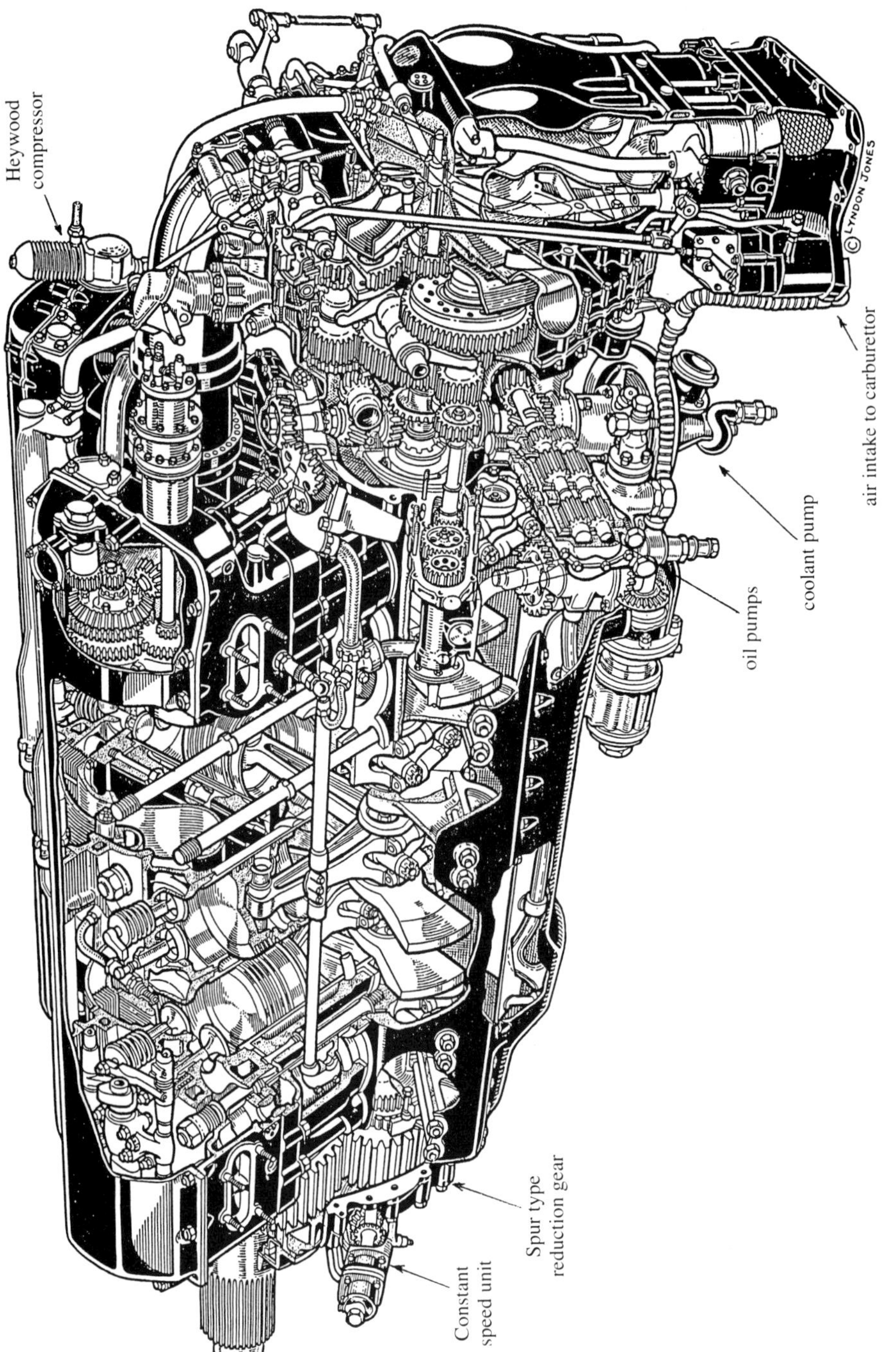

Heywood compressor

air intake to carburettor

coolant pump

oil pumps

Spur type reduction gear

Constant speed unit

© LYNDON JONES

Figure 10. Rolls-Royce Merlin XX engine.

G-AFBT. The main work concerned sleeve wear around the ports, producing a reliable reduction gear and the development of a strong sleeve crank. With the introduction of the Mk. VI came a number of other improvements, Salomen vibration dampers in the crankshaft balance-weights, closer pitch fins on the cylinders for improved cooling, as well as the first use of 100 octane fuel. This was the engine originally planned for the Beaufighter.

With the Hercules standing stationary for a time it is always necessary to turn the engine over by hand, to ensure that oil has not drained down to the bottom cylinders, where the piston can come in contact with it during the compression stroke (hydraulicing) and so cause breakage and expensive engine noises! With the Hercules VI there was a separate mixture control, but this was incorporated into a single lever control with the Mk. XVI and other marks.

The second engine used on the production Beaufighter was the Rolls-Royce Merlin. This had its origins with the development of the Kestrel, through the Buzzard and its development into the 'R' engine; for with the 'R' were introduced higher BMEPs and piston speeds. The basis of the Merlin was the scaling up of the Kestrel with the displacement increased from 1295 to 1649 cubic inches.

Detail design began in 1933, with the first drawings being issued to the workshops in April. By October the first two experimental engines were on test, with the cylinder block and upper crankcase being produced as a single casting. Up to this point the total cost was borne by Rolls-Royce, and the engine was known within the Works as the PV.12. Further development work resulted in a redesign of the reduction gear and the strengthening of the block casting; this was followed by work being commenced on another type of cylinder head, as the test results indicated a need for improvement to give a shortened flame travel and a high degree of turbulence.

This latter development was named the 'ramphead' and was first introduced on the PV.12 'B' model and used on all engines up to the 'F' model. The bench tests however soon revealed serious problems, so a scaled-up Kestrel head was introduced and installed on the 'G' model, which became the Merlin II. Further development brought separate cylinder block and crankcase castings, and numerous other small improvements.

Then in January 1935 work began on development of a two-speed drive for the supercharger, but difficulties were experienced with the drive, and so Rolls-Royce licence produced a Farman drive. A Merlin with this type of drive and two-speed supercharger was first flown in September 1937, and the engine went into production as the Merlin X. This work was all done under J.E. Ellor, followed by work on the supercharger being carried out by S.G. Hooker, which resulted in an improvement in supercharger efficiency; this engine went into production as the Merlin Mk. XX, the engine chosen to power the Beaufighter II.

The Merlin was a conventional poppet-valve engine of Vee cylinder block configuration of twelve cylinders with liquid cooling, using 70/30 water/Ethylene Glycol in a pressurized coolant system. The Merlin engine on the Beaufighter offered very little difference in flight performance, but allowed greater aircraft production.

Aircraft handling

It would appear that initially at OTUs some pupils gained the erroneous impression that the Beaufighter was a deathtrap, which was certainly not the experience of squadron personnel. Some of this may have been due to the fact that often 'rogue' or 'tired' aircraft were routed to OTUs, and at that period, experienced pilots were required by the squadrons and not working as instructors; once recognized the situation was rectified.

A lot of nonsense has been written over the years about the take-off swing of the Beaufighter, which was in fact no worse than that of the Mosquito, and could easily be held. It is obvious that any aircraft whose CG is well behind the main wheels, with two powerful engines forward, is a natural for moving from the straight and narrow as soon as the engines are opened up. This fault on the Beaufighter was easily corrected by some differential throttle control when opening up, maintaining the aircraft in the forward direction until the rudder 'bit' and full rudder control was gained — then moving the throttles to the gate and the Beau accelerated like a scalded cat.

Cockpit layout of R2055 prototype, before operational equipment cluttered it up!

From the cockpit the forward view was excellent, probably the best ever, and the wingtips could also be seen; but a first glance into the cockpit interior initially gave the impression of untidiness, the only exception being the standard blind-flying panel. Yet strangely enough, for all its apparent untidiness, the cockpit and its switches and gauges seemed to fall readily to hand — even if sitting 'up front' did give one the impression of playing the Mighty Wurlitzer!

* On take-off align the aircraft on the runway and ensure the tailwheel is straight; open up the engines against the brakes to check both engines respond evenly. In opening up the engines, do so slowly and evenly, keeping straight by coarse use of the rudder, then raise the tail early to improve the rudder control. Safety speed at full load (including torpedo) at the full take-off power, with flaps 'up' or 15° down, is 198 mph IAS.

After undercarriage up, and at not less than 300 ft retract the flaps, then select power off. The recommended climbing speed is 172 mph IAS and for maximum rate of climb 150 mph IAS.

The elevator and elevator trim controls are light and powerful and must be used with care; while the stalling speed varies with aircraft fully loaded with or without external load. With the aircraft 'clean' the stalling speed is 110 mph with external load and 104 mph with no external load. With the undercarriage and flaps 'down' the stalling speed is 86 mph fully loaded with external load and 80 mph without external load.

General flying at normal loads has no problems, as the aircraft is stable about all axes; but as the aircraft approaches the stall there is slight elevator buffeting, then at the stall the nose drops gently. In a steep turn, elevator buffeting occurs just before the stall, at which the wing may drop sharply.

---

* Applicable only to Hercules-powered versions.

Cockpit layout of operational Beaufighter TF.Mk X.
(B.Ae Bristol)

On the approach to land, turn into wind at approximately 120 knots (138 mph), and progressively reduce airspeed so as to cross the threshold at 95 knots (109 mph) with the flaps down. The aircraft is easy to land on three points if power is not reduced suddenly; the elevators however are light, so care must be taken not to 'balloon', but once the tail is on the ground the aircraft decelerates quickly. Beaufighters without the dihedral tailplane are unstable in pitch, especially with flaps 'down'.

The modified tail aircraft is longitudinally stable in level flight down to 140 mph, and on the climb neutrally stable at the speed for maximum climb (140 mph from SL to 18,000 ft). On the glide the aircraft is still stable, both with flaps and undercarriage, up or down. The elevator trim tab control is designed to assist manoeuvring and recovery from dives, but must be used with care and slowly as it is powerful, so that sudden use could result in heavy stresses being placed on the airframe.

The Beaufighter's great structural strength allowed the maximum speed for undercarriage down to be as high as 240 mph IAS (the later Mosquito FB6 was only 178 mph), so undercarriage lowering was one method used by Beaufighter pilots, to reduce the forward speed on overtaking their quarry, while at the same time maintaining sufficient power on their Hercules engines for any emergency or manoeuvring.

Chapter 5
In Service

Introduction

At the beginning of June 1940 the Air Staff were urgently requesting that a Beaufighter Squadron was formed for operational duties, but the first eleven Beaufighters had been allocated to various test establishments or for trial installations as per the test programme. On the 5th, four of the aircraft were reallocated for issue to Service units, followed on 7 July by the Air Ministry deciding that the first unit to be equipped with the Beaufighter would be 25 Squadron. The first two aircraft delivered were delivered less guns and gunsight, but orders were given that all future deliveries were to be fully equipped. The normal procedure of a Service Squadron carrying out an intensive 100 hours flying was abandoned, but the teething problems

The clean and pugnacious line of a Mk I Beaufighter in daylight colours 1940. (F. Coombs)

had not been corrected, and the abandonment of the programme would result in a large number of aircraft unserviceable during late 1940.

With the end of the Battle of Britain and the commencement of the night Blitz, the need for a night interceptor became a top priority; for in spite of various past specifications calling for a day and night fighter, without radar none of these could perform an adequate function in the night fighter role. The Beaufighter however had sufficient performance for the role and sufficient space to accommodate AI radar, and it was now entering service; furthermore, its forward-firing armament was the heaviest of any fighter. On 26 July 1940 the Beaufighter was cleared for RAF service, and the decision made regarding the re-equipment of fighter and strike-fighter squadrons in the proportion of two in Fighter Command and one in Coastal Command; but on 23 September this decision was rescinded, and the order was

given that the first 100 aircraft were to be delivered to Fighter Command. A month later the Air Ministry stated that the Blenheim 1F squadrons of Fighter Command were to be re-equipped with Beaufighters, with the exception of 23 Squadron.

1941 commenced with the Air Ministry on 6 January notifying both Fighter Command and Coastal Command that the production of Beaufighter aircraft would be as follows:

(a) Beaufighter 1 from Bristol present contract to Coastal

(b) Beaufighter 1 from Fairey Stockport contract to Coastal

(c) Beaufighter 2 to Fighter

(d) Beaufighter 1 from Bristol (Weston factory) to Fighter

Apart from the teething troubles with the airframe, trouble was being experienced with the 20 mm cannon, whilst the framing of the pilot's cockpit and windscreen was thick with small panels, which restricted the visibility — a feature that was not helpful for a night fighter. As a night fighter the Beaufighter had much to recommend it, not least being the fact that it was the most heavily armed fighter on either side at that stage in the war. It was also well armoured and was reasonably fast for the time, as well as being a steady gun platform. It also had a reasonable endurance in standard form, which lent itself to the standing patrol system that British night fighters used in 1941–42. The Beaufighter was to be developed into one of the most potent weapon carriers of the 1939–45 war, and would be used in every theatre of operations in many roles; it became a long-range fighter, night fighter, intruder, torpedo fighter, fighter bomber, ground attack and strike fighter. Not only being developed under the auspices of Bristol, RAE and A&AEE, but also in the field with local modifications suitable to the theatre of operations.

The demand for the Beaufighter was such, that the Air Staff as well as other officials pushed for its entry into service, and for an ever increasing supply; this is reflected time and again in correspondence between Fighter Command to MAP and the Air Ministry. Although a Blenheim of FIU flown by a specialist crew (Flying Officer Ashfield, Pilot Officer Morris and Sergeant Leyland), had on 22 July 1940 carried out the first successful radar interception and destruction of an EA (Dornier 17), the Blenheim was insufficiently fast enough in the interception role. Thus the Beaufighter was required to rectify this situation. Unfortunately defects in the cannon meant that nearly all the Beaufighters were unserviceable due to armament; similar defects affected all other cannon-armed aircraft. In the case of the Beaufighter the defects were further complicated by incorrect assembly of the cannon mounting at Bristol's. Eventually a new design of mounting with a swing link was introduced and stricter inspection authorized.

After the Beaufighter was cleared for issue to the RAF on 26 July 1940, five were dispatched the following day from Filton to the MUs (Maintenance Units), where they were equipped with certain installations. The first four aircraft issued going one to each of four squadrons, R2056 to 25 Squadron, R2070 to 219 Squadron, R2072 to 29 Squadron and R2073 to 604 Squadron, starting from 2 September 1940.

The first operators of the radar on the Blenheims and Beaufighters were the first in service, and in the case of the Beaufighter were often ex-air gunners from Defiants or Blenheims, who had stayed on to fly as gun loaders on the Beaufighters when they arrived — no radar being fitted. Then with the installation of radar, those who had an inclination to dabble with it, were given local training by radar operators and radar mechanics. The radar operator, when fully trained, received a half wing enclosing the letters RO, though their official title was 'Operator (radio)'. In Fighter Command the second crew member's role was radar operator, gun loader and lookout, and he was issued with a seat-type parachute — though some squadrons may of course have deviated from this rule with respect to the parachute type. Gun loading was of course the loading of ammunition drums before belt feed was introduced; each ammo drum held sixty rounds of 20 mm ammunition and was very heavy, so changing drums at 20,000 ft during a combat was some feat — so to encourage fitness and competence some squadrons, such as 25 Squadron, held competitions on the ground.

When the Beaufighter and its radar was introduced into Coastal Command the role of the second crew member was a little more complicated, being the same as that in Fighter

Beaufighter VIF V8565 of the F.I.U. at Tangmere, fitted with AI.Mk VIII radar.

Command plus the duties of navigator. The aircraft was equipped with a short navigation table, navigation equipment and DF radio; so to give the RO a greater freedom of movement he was issued with a lap-type parachute, which could be stowed. The radar operators were given navigation courses and then listed as 'Navigator (radio)', and could then wear the 'N' half-wing badge; some preferred to retain their RO badge. In Fighter Command in about 1942, some of their ROs went on navigation courses, and they also were allowed to wear the 'N' badge. On North Sea sweeps, before the installation of the VGO or Browning 'scare' gun, some ROs carried Thompson sub-machine-guns, described by one '. . . as some sort of defence', but this of course necessitated the smashing of the RO's canopy before the gun could be used.

Ken Lusty* became a Navigator (radio) and says of those times of change

> In early 1940 I was an air gunner on 25 Squadron which flew the fighter version of the Blenheim. When we heard that the Blenheim was to be replaced by Beaufighters we gunners were very disturbed — we heard that there was no place for a gunner on the Beau'. The Beaufighters had six forward-firing 0.303

inch Brownings and four 20 mm cannon. The Brownings were in the wings — belt fed, and the cannon were under the nose but the guns themselves were in the body of the plane and were *not* belt fed. They had very heavy drums of 20 mm shells which had to be replaced after firing. We were to be POWDER MONKEYS!!

> Well, not quite. After very little training we re-mustered as Radio Navigators. Changing those heavy drums of 20 mm shells was very difficult in total darkness at night and many a finger — or two — was bruised in the process.

The Fighter Interception Unit (FIU) received R2055 from the A&AEE on 12 August, followed by R2059 on 1 September; this latter aircraft having the distinction of carrying out the first Beaufighter operational night sortie three nights later. Within the week 600 Squadron had received their first Beaufighter and Squadron Leader M. Maxwell had been appointed their Commanding Officer. On 30 September the squadron flew its first operational sortie, but recorded no 'kills'.

At the start of October the first complaints were being received at Bristol's from Fighter Command, amongst which were the following:

---

* Left the RAF at the end of the war as Squadron Leader DFC.

bullet-proof windscreens cracking, cabin air intake on engine cowling cracking, no holes in engine cowling for starting handles, and aileron control cables fouling. This was followed on the 5th of the month with a meeting at Fighter Command, where the AOC-in-C expressed his dissatisfaction with the speed of the Beaufighter (obviously ignoring the fact that the aircraft were still powered by the Hercules III and not the VI). Three suggestions were made, use boost override and 100-octane fuel in the engines, get rid of the hot-water boiler for cabin heating (which he understood absorbed 60 hp per engine) and streamline the aerials and generally to clean up the airframe.

It would be fair comment to say that the performance of the fully-equipped R2054, with such a reduced maximum speed from the specification, was not only a low point in the Beaufighter's career, but also a disappointment for the Bristol Aeroplane Company, and feverish activity took place within the works to rectify it — which was not exactly helped when the AI radar equipment weight and aerials were added. Apart from the fitment of the Hercules VI engines, which Bristol were trying hard to clear for production, they did make the suggestion to use the Hercules in a close-fitting, low-drag engine cowling that employed reverse flow cooling. If this alteration had been adopted it would have taken even longer to develop and put into production than that involving the Hercules VI. In spite of rumblings at MAP and statements by some of its members, it must be realized that Bristol were aware of the potential of their sleeve-valve engines, and that a firm at Derby were their rivals in the aero-engine market, so therefore all haste was required to clear the Hercules VI and its successors.

The modification programme to equip the Beaufighter with AI Mk. IV radar was also not yet on a production basis. The aircraft to carry AI radar, once cleared from their production centre, had to be flown to 32 MU St Athan for the installation of this highly secret piece of equipment — this programme only commenced from September 1940. After that the aircraft was cleared for flight to one or other of the MUs concerned with the preparation or storage of Beaufighters, or direct to the squadron, if its preparation was complete.

Air Interception radar was new, the Beaufighter was an interim design pushed into production, yet, as the Hurricanes and Spitfires defended in daylight the unconquered British people in their island kingdom, so would the Beaufighters guard the night skies against the German invaders — then strike by day and night at the enemy's ships and airfields, both in Europe and on the seas from Norway to Spain, in the Mediterranean, over the North African desert and against the Japanese in the Far East. It also gave to the RAAF the weapon to carry on the fight, and to strike at any Japanese stronghold, airfield or ship.

Night fighters

The Fighter Interception Unit (FIU) of the RAF came into being to develop interception techniques in conjunction with radar, and was formed in April 1940 under the command of Squadron Leader Peter Chamberlain. It was based at RAF Tangmere and initially equipped with Blenheim fighters. The unit's first Beaufighter was R2055, which arrived on 12 August and was fitted with AI.Mk. IV radar, however, Tangmere suffered from a daylight attack and the aircraft was slightly damaged, so the unit made a temporary move to Shoreham.

This delayed the Beaufighter's first operational sortie until 4/5 September, when 'Jumbo' Ashfield and crew flew around but had 'no trade'. Further flights were made on other nights, but the aircraft failed to return on the night of 12/13 September when flown by Flight Lieutenant Ker-Ramsey and crew; it went down in the sea off Boulogne.

The Beaufighter did not open its score until the night of 25 October, with R2097 of 219 Squadron crewed by Sergeants Hodgekinson and Benn; but this was without radar and the target was a Do17. It was around this period that AI.Mk. IV radar was installed in squadron aircraft, and the aerials of this equipment could be identified by their 'arrowhead' appearance and were fitted on the nose of the aircraft (see Figure 3 of Beaufighter Mk II).

With this equipment, on the night of 19/20 November, Flight Lieutenant J. Cunningham*

* Left the RAF as Group Captain J. Cunningham DSO DFC, and later renowned for his post-war test flying for de Havilland.

and Sergeant J. Phillipson flying R2098 of 604 Squadron, scored the first radar-assisted victory in a Beaufighter — exactly five days after the Blitz of Coventry, when the RAF fighter force was shown to be totally inadequate. The Beaufighter and its crews were at this period in a state of experimentation, of learning — and of frustration; still learning the intricacies of the 'magic box', trying to gain experience, struggling to evaluate its indications in relation to their aircraft and the target — all surrounded by the sheer blackness outside and possible cannon malfunction inside. 219 Squadron for instance, had nearly all its Beaufighters grounded with cannon defects. Cannon malfunctions in the air required the RO to remove the heavy magazine, with the gunbay in semi-darkness, and the aircraft swaying and rolling in flight — no easy task.

In October 1940 Blenheims and Beaufighters were being used by 25 Squadron for night fighting, the first Beaufighters having been received from 27 MU in late October. Squadron Leader H. Pleasance was posted in from 12 Group Headquarters to take command of the squadron, but in 1941 he was posted away and the command passed to Wing Commander Atcherley. Meanwhile the squadron lost their first aircraft and crew on 23 November, when Pilot Officer Marsh and his RO were killed when they crashed into the sea after take-off from Tangmere. By the start of December some more Beaufighters had been received, but then on the 8th of the month, six of their Beaufighters were turned over to 604 Squadron.

On the night of 23 December the *Luftwaffe* pathfinder unit, KGr100, were to lose their first aircraft to a Beaufighter. This was a He.111 codemarked 6N+DL, which was attacked by Cunningham of 604 Squadron. The Beaufighter was brought into position by gun-laying radar, and the He.111 was attacked at approximately 16,000 feet over the Channel and crashed near Cherbourg. The Beaufighter team of Cunningham and Phillipson were already proving themselves an 'ace' crew, for the night previous to this combat they had destroyed a Ju88.

In spite of interception patrols no EA were seen by 25 Squadron during the early months of 1941, but the squadron lost their first aircraft of the year when R2215 crashed on landing on 29

March. It was in March when night fighter claims of enemy 'kills' started to improve, but it was not until 9 April before 25 Squadron claimed their first Beaufighter victim, when Sergeants Bennett and Curtis in R2122 intercepted and shot down a Ju.88.

Weather over the winter months had given no help to operations, and even 'starred crews' — expected to take-off in any weather — were short of targets. So it was early April 1941 before prospects improved; and during the month twenty-two enemy aircraft were shot down; 604 Squadron operations book recorded that on 4 April six patrols were flown and three EA claimed as shot down. Unfortunately the weather closed in, resulting in the loss of a Beaufighter, when Flight Lieutenant Lawton/Sergeant Patson could not land and were forced to bale out, which they did successfully.

At this period there was in being a radar landing device known as 'Father', but it was considered by the crews as a bit hit or miss. This device was installed only at certain airfields, so the crews not only had to know which airfields, but also their condition. This is more understandable when one realizes that the Beaufighter was a heavy aircraft for the time, weighing the best part of 10 tons; so that landing it downhill required quite a lot of room, whilst landing it on wet grass could result in wheel-lock and an aircraft merrily skidding on — and at night who knew what was on the ground in front.

The night of 9/10 April found KGr100 losing another He.111 to 604 Squadron Beaufighters. Flying Officer Chisholm and RO Sergeant Ripley took off from their base at Middle Wallop, and under the guidance of GCI were vectored onto He.111 6N+BK. Visual contact was made at about 10,000 feet above Ringwood, and a blast of cannon dispatched the EA. 604 Squadron soon became the lead scorer in shooting down night raiders, but during the next few months of 1941 other squadrons started to have their 'ace' crews; Wing Commander Atcherley of 25 Squadron on 5 May commenced his Beaufighter score, when, with his RO Flight Lieutenant Hunter-Todd, they shot down a Ju.88 near Bourne.

A diminishing *Luftwaffe* night offensive during summer 1941 was due to preparations for the launch of 'Operation Barbarosa' as well as the British night fighter force becoming more

604 Squadron night-fighter IF T4638 with its tailwheel locked 'down'.

effective. For by the start of 1941 came the linking up of the AI-equipped Beaufighter, with the GCI (Ground Controlled Interception) system, and with the radar chain. Thus began the changeover from comparative failure in the winter of 1940 to a successful start in the spring of 1941, as both the aircrew and the GCI controllers gained expertise in the operation of the overall system — and the *Luftwaffe* obligingly provided targets.

By the February six squadrons had or were, converted to Beaufighters; these were 25, 29, 68, 219, 600 and 604; the rest of the night-fighter units were flying a wide selection of other types of aircraft, but the only effective night fighter was the Beau. This is confirmed by a letter from Air Marshal W. Sholto Douglas to the under-Secretary of State for Air on 8 April. In this letter Douglas indicated the state of Beaufighter supply and wastage, as well as the score card for Beaufighters and other types of night fighters; with an assessment that:

> . . . although Beaufighters fitted with AI carried out only 21 per cent of the sorties at night, they have been responsible for 65 per cent of the enemy aircraft destroyed . . . I most strongly urge that steps should be taken to increase substantially at an early date the supply of Beaufighters to Fighter Command.

Although the Prime Minister was initially in favour of diverting some Beaufighters to Coastal Command, with the successes in night fighting achieved with the Beaufighters, Churchill, at a Battle of the Atlantic conference in the middle of April, stated that he wanted as many Beaufighters as possible employed as night fighters. At this stage there was only one Beaufighter squadron in Coastal Command, yet there was almost as much need for them in that Command as in Fighter Command, but obviously the defence of the home base took priority. On 18 May Douglas had written to Freeman about Beaufighter supply, again mentioning the six squadrons that were equipped with these aircraft: 'a strength of eighty-one Mk. I aircraft and five Mk. II aircraft', quoting the shortage of Beaufighters and that he was activating 406 Squadron at Acklington with Beaufighters — in other words 'give me more'.

The Beaufighter was now in demand everywhere, but production was still a problem; it was estimated by MAP that by June Bristol's would have increased production to eighty fighter aircraft each month, and that Fairey would be producing thirty coastal aircraft each month.

The handling of the Beaufighter has been described a number of times, by test pilots and delivery pilots, most of whom considered it a robust aircraft, although the swing on take-off needed watching, but a safe aircraft if handled correctly. The Hercules-powered Beaufighter needed to be flown in the prescribed manner, to be mastered and no deviations made. Though heavy on the ailerons and far from a manoeuvrable aircraft, its strong construction bred confidence in its crews. Group Captain John Wray says of it:

> The manoeuvrability of the Beau was variable and depended on its rate of roll. Its size and weight did not lend it to fighter type manoeuvres, and in any case, the tactics we employed (at that period) did not necessarily require great manoeuvrability. On most aircraft the rate of roll was fairly slow, the ailerons being rather heavy. However, there was the odd aircraft that had a very much better rate of roll, and this was noticeably much more pleasant. I can only assume it must have been something to do with where the airframe was made.

Flying Officer J. Wray when flying Blenheim of 53 Squadron 1940. (G/Capt J. Wray)

The single-engined performance was satisfactory, although of course we had no feathering on the airscrews.

This was in 1941 and early 1942, when the German bombers tended to fly straight and level on their approaches and retirement from the target area, so the tactics were designed to counter this. At that period the country in the south and south-east was ringed with large CH (radar) stations giving a 'floodlight' view of the areas out to about two hundred miles, depending on the height of the target aircraft. These were the early warning stations, but could not give a target's height or speed, so were supplemented by the CHL and CHEL stations, the latter having the capability of seeing down to sea-level. All the stations fed into the control system, where the information was filtered and passed to the Group and Sector Operational Rooms. The CHL and CHEL were also used to control the night fighters, directly controlling them until an interception took place. When the target aircraft was picked up, the Beaufighter night fighter was vectored onto it and closed on the target, then synchronized speed with it, making an approach from the rear. Once the pilot had a visual sight, he aligned with the target, got in range and opened fire — so there were few violent turns

BEAUFIGHTER II
MERLIN XX
SEPT/41

Mk II night-fighter R2402 of 255 Squadron in 1941.
Note the enclosed exhaust of the Merlin XX engines.

and manoeuvres, unless the fighter missed and his target took evasive action.

On the night of 15 April 1941, J. Cunningham and C. Rawnsley flying Beaufighter R2101 of 604 Squadron, operating from Middle Wallop, scored a hat-trick on night interception; first stalking and shooting down a He.111 — with the help of good controlling, Army searchlights and AI radar. Then on the second sortie freelancing from base, Cunningham spotted searchlights over in the Marlborough direction, and nosing the aircraft down with throttles forward, a visual contact was made going over Southampton. Closing the gap to approximately 80 yards Cunningham fired and the He.111 received a full blast of cannon fire, and nosed over into a dive to smash into the ground. The night was not yet over, for upon returning to the Middle Wallop circuit, Cunningham was vectored on to an EA heading homewards fast. Banking their Beaufighter onto the new heading with the He.111 now approaching the outskirts of Southampton, the RO got a contact, again it was below their height. Cunningham eased R2101 down slowly closing the gap, then at 80 yards visual the aircraft was identified and was shining silhouetted in the moonlight — a blast of cannon and their third victim exploded into burning wreckage.

By 10 May Filton had dispatched its 200th production Beaufighter. These were needed for the 'Night Blitz' on London on 19/20 May, when Beaufighters claimed over half of the twenty-four enemy bombers shot down. 406 (Lynx) Squadron RCAF had by this date been formed at Acklington with Blenheims, to be the first RCAF night-fighter unit in Europe. They gradually converted on to the Beaufighters, but flew a mixture of Blenheims and Beaufighters during June. It was at this period of forming and training commanded by Wing Commander D. Morris DFC.

During July another unit commenced converting on to Beaufighters — 255 Squadron commanded by Wing Commander J.S. Bartlett DFC. The unit had been patrolling the night skies in Hurricanes and Defiants, and now began to receive Beaufighter Mk. IIs. This unit was followed by 141 Squadron at Ayr and 409 (Nighthawk) Squadron RCAF based at Coleby Grange receiving Beaufighters, which, in the case of 409 Squadron were the Mk. II version. Command of the squadron passed to Wing Commander P.Y. Davoud RCAF, who was also responsible for the squadron's first EA victory.

On 1 November 1941 with RO Sergeant Carpenter, Davoud took off from Coleby Grange at 2055 hours and given several vectors by the Orby GCI, obtained a contact on their AI at a maximum range of 11,000 ft, with the target well to port and 500 ft below. Davoud increased their speed and turned to port, obtaining a visual at 6000 ft with the EA silhouetted against the cloud in bright moonlight. He throttled back and lost height, dropping 400 yards to the rear of the EA which then observed the Beau and made a dive for the clouds . . .

> I closed to approximately 200 yards, identified bandit as Dornier 217 and fired short burst, observing hits on starboard mainplane. The Dornier returned fire, and having closed to about 100 yards I fired two long bursts, seeing the second burst hitting his starboard engine. Just before Dornier entered cloud, a big explosion blew his right engine and wing off. I pulled up to avoid a collision and the Dornier fell burning straight into the sea.

During this engagement, one of the 20 mm cannon had refused to fire.

Occasionally the *Luftwaffe* flew intruder patrols, but fortunately these were spasmodic, for Bomber Command as well as Fighter Command would have been embarrassed. One of these intruders caught one 25 Squadron crew on the hop after a long night sortie — Ken Lusty explains:

> The Beau was built like a tank — it could soak up punishment and still fly. We were on the approach at Wittering one night after a rather long patrol and probably not as wide awake as I should have been. A Ju.88 hit us with over one hundred rounds, but we still managed to get down.

Air fighting at night was by now considerably better for the British night-fighter, as AI Mk. IV coupled with CH. CHL, CHEL and IFF, plus the VHF replacing the HF communication sets, were now making 'Flash your weapon'* a more potent and viable command (ROs and Nav' ROs, in correspondence felt that AI Mk. IV was not too bad, even when compared against later sets). Night fighting though, could never be compared with day fighting, for it lacked the dash; it was slower, more devious, more scientific. The pilot and RO worked as a team, but no matter how good the RO was, in the final analysis it was the pilot who had to see the target and administer the *coup de grâce* .. yet, there was the need for a good RO to get the pilot to that position. Initially the 50 mph sight was used for night fighting, as

the attack was usually delivered from the rear with very little movement in evasion of the target; but when the *Luftwaffe* introduced more manoeuvrable and faster aircraft such as the Ju.88 and Dornier 217, attacks had to be made from an angle, and then manoeuvrability of the 'hunter' was at a premium. To cope with the greater deflection angles of aiming, the GM2 gunsight was introduced.

At this stage in the Beaufighter story it may be worthwhile to bring in to the picture an officer who had all the aggressiveness that a born leader needed, Group Captain Basil Embry. Embry had led 107 Squadron Blenheims in attacks on German forces during the campaign in France in 1940, had been shot down and escaped, then accepted command of a night fighter wing in 11 Group and reverted to the rank of Wing Commander. The Germans were to regret over the years the escape of Embry, for whatever or wherever he commanded, his idea was to lead from the front — and that meant a cockpit. Now he took command of the Wittering sector, which included the parent station of Wittering and the satellite airfields of Coleyweston and Kings Cliffe. So it was not long before he was to be found in the cockpit of a Beaufighter, which was serviced by 25 Squadron; however, he had no regular AI operator. This was soon rectified when he acquired Peter Clapham, who was a fighter controller in the sector operations room and was medically unfit for flying duties! No wonder the Wittering sector was known as Embry's Air Force, both jocularly and with pride.

The two began a partnership that lasted the war years, and the team flew regularly on sector patrols with 25 and 151 Squadrons. These sector patrols meant maintaining two or three aircraft on continuous patrol, and positioning them in relation to the GCI field of 'view', so that the controller could immediately direct the nightfighter to the EA.

On the night of 14 June 1941 Wing Commander D. Atcherley of 25 Squadron flying R2251 shot down a He.111 near Sheringham, whilst another pilot of the same squadron caught a Ju.88C, code marked R4+DM, of 4/NJG2 over the UK and shot it down near Narborough. 219 Squadron was not to be left out of the shooting

---

* Command from GCI controller to RO to use his AI radar.

stakes, so their Flight Lieutenant Topham took off from Tangmere, chased and caught a He.111 (code marked 6N+FK) of KGr100 and shot it down over Dorset.

Then on the morning of the 17th, R2251 of 25 Squadron took off from Wittering at 0134 hours crewed by Wing Commander Atcherley and radar operator Flight Lieutenant Hunter-Todd on another interception. By 0245 hours Orby GCI had vectored them on to a Ju.88 travelling out to sea in an easterly direction; within ten minutes Orby had ordered the crew to 'flash your weapon'. One minute later Hunter-Todd obtained a contact on an EA about 10,000 ft ahead at an altitude of 10,000 ft, and so:

> After about 3–4 minutes I got a visual of the EA 600 ft ahead, slightly left and below, and I identified it as a Ju.88. Closing in I opened fire from dead astern at about 100–150 yard range and after a two second burst the EA blew apart in the air, fragments hitting the port mainplane and fuselage of the Beaufighter.

Once on the ground a physical examination of the Beaufighter turned out a little gruesome, for as well as there being two feet of the leading edge of the port wing stoved in, there were also unmistakable traces of human remains on the port wing and spots of blood on the port engine cowling and propeller.

The end of September 1941 found Fighter Command equipped with twelve Beaufighter squadrons: 25, 29, 68, 141, 219, 255, 307, 406, 409, 456, 600 and 604, 456 Squadron being the last to equip. 141 Squadron at Ayr had received their Beaufighter 1Fs in August, as had 409 RCAF Squadron at Coleby Grange. In the same month 307 (Polish) Squadron commanded by Squadron Leader J. Antonowicz and based at Exeter commenced re-equipping with Beaufighters.

307 Squadron had originally been formed at Kirton-in-Lindsey on 5 September 1940, and within two days received Battle R7411 and Master N8009 for training purposes. The squadron was however destined to fly the night skies of Britain in Defiants. Then on 3 August 1941 a letter from HQ 10 Group advised the unit that they were shortly to receive Beaufighter Mk. 1F aircraft. The following day two British LAC Observers (radio) were posted in and Pilot Officer Kownacki advised he was posted to 307 Squadron for the training of AI operators.

The type of Beaufighter to be supplied to the Squadron was changed on 8 August, when they were notified that they would receive Merlin-powered Mk. IIs. Six days later the aircraft started rolling in, when R2443, R2446, R2447 and R2449 arrived from 19 MU. Over the next two weeks a further twelve Beaufighter Mk. IIs arrived from the MUs, and flight training was enthusiastically entered into by the Polish aircrew; flying being carried out on Blenheim and Beaufighter aircraft. Then on the 30th of the month they suffered the first loss of one of the Beaufighters, when R2315 had its port engine fail immediately after take-off; the aircraft crashed and burnt out, but the crew, Sergeants Malinowski and Modro, were only injured.

The squadron suffered its first loss of Beaufighter aircrew on 27 September when R2442 flown by Pilot Officer Gayzler and Pilot Officer Pfleger flew into a hill at Hameldown Moor and the two men were killed. By 15 October three crews had been declared as up to operational standard, and on the same day at 1855 hours the first operational interception sortie was flown by Squadron Leader Antonowicz and Flying Officer Karwowski in R2463.

Although interception patrols were flown from this date, practice flying continued on both the Blenheim and Beaufighter, but it was not until 1 November that the first combat took place. At 1950 hours R3019 took off, flown by Sergeants Jankowiak and Karais, contact was made and the pilot sighted a Ju.88 at 14,000 ft. The pilot closed to 200 yards and gave a 4-second burst, then continued closing to 50 yards, when he gave a 2-second burst of fire, at which the EA dived steeply to port and contact was lost.

This was followed by R2379, flown by Sergeants Turzanski and Ostrowski taking off at 2140 hours and obtaining a contact on an EA, followed shortly afterwards by a visual. The EA was identified as a Do.217 and sighted 200 yards ahead and about 300 ft above. The pilot gave a 3-second burst into the EA's port engine, which burst into flames, then a further 2-second burst into the fuselage, when pieces were seen to fall off. A further 2-second burst of cannon was then poured into the port wing, which became detached in a large explosion and the aircraft

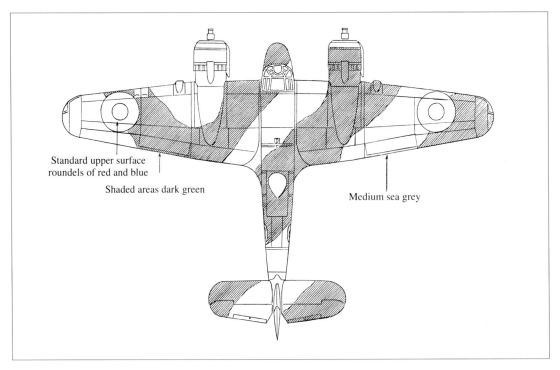

Figure 11. Above view of Beaufighter night-fighter paint scheme after later 1942.

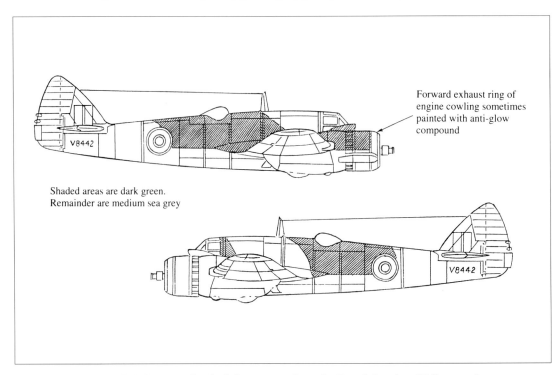

Figure 12. Side views of night-fighter paint scheme for Beaufighter late 1942 onwards.

crashed in flames; this was confirmed by a searchlight post. A second AI contact was made at 2318 hours and the aircraft recognized as another Dornier; the EA immediately commenced evasive action and dived to 8000 ft, closely followed by the Beaufighter. One burst of cannon was made from 150 yards and closing to 50 yards, resulted in a large explosion and dense smoke; the EA then dived vertically in flames and was claimed as destroyed.

Another unit equipping with the Beaufighter II was 410 (Cougar) RCAF Squadron, which like 307 Squadron, had been formed on Defiants. Then in April 1942, with the squadron stationed at Drem and commanded by Wing Commander C. Lipton, re-equipment began with Beaufighters, the first two T3387 and T3152 arriving on the 2nd, followed by eleven newly trained ROs from 3 Radio School arriving on the 21st. Meantime flying continued on the Defiants, and twin-engine training commenced on a Blenheim, as more Beaufighters arrived.

By 7 May the squadron had nineteen Beau's on charge, but the same day lost their first one, when the pilot of T3374 was forced to crash land as one undercarriage unit would not lower and lock down by any means. The following day two crews were classed as operational, although officially the squadron was still non-operational. The enthusiasm of the squadron to get operational was noted in a letter to the squadron from the Station Commanding Officer, and on 5 June three Beaufighters scrambled on an interception, but made no contact.

The 16 June found the squadron starting another move, this time back to Ayr, although four aircraft and five crews were to remain at Drem. Then on 5 July the squadron suffered its first serious crash, but without loss of life, when V8132 flown by Pilot Officer Devlin had both engines cut at 2000 ft. The port one caught fire, so the pilot cut all switches and glided (!) straight ahead, clouting a power pole before crash-landing in a field, killing three cows, but the crew jumped out safely.

During the month of August a number of attempted interceptions were made but without success, then at the end of August the whole squadron moved to Scorton and Wing Commander Hillock took over as CO. Following this, on 7 September, the squadron notched up its first action on the Beau; T3428 flown by Pilot

Officers R. Ferguson and D. Creed were on a GCI exercise, when they were vectored on to an EA. Contact was made, then a visual, which identified it as a Ju.88. Closing in and opening fire, hits were seen and the EA dived away, so it was only claimed as damaged.

By 1942 there were doubts being expressed about the effectiveness of the night fighters' black camouflage; one leading advocate for changing the colour scheme was Max Aitken, the Commanding Officer of 68 Squadron at Coltishall. A series of proposed colour schemes were put forward and 68 Squadron authorized to test them out; this resulted in visitors to Coltishall seeing Beaufighters with stripes, stars all over them, a pink one and other various schemes and colours. The Fighter Interception Unit and 85 Squadron also tried other schemes, the former trying out an overall dark green.

Twelve colour schemes were also tried out at RAE Farnborough in their laboratory, which indicated that light colours were not satisfactory, and that black was one of the best finishes. However, operational experience had already proven this incorrect, and the aim was now to find a compromise colour that would be satisfactory on dark nights, not conspicuous in moonlight or in searchlights, and provide satisfactory day camouflage on airfields. All of these trials and tests eventually settled on an overall medium sea grey and a dark green in disruptive pattern. The roundels were also changed, with the upper surface ones being a Type B of red and blue, with no under wing roundels, and the fin flash conforming to the standard 1942 version, where the white was only two inches wide. The new colour scheme was authorized in October 1942 with production aircraft being painted to follow it in about November.

John Wray was at that period taking delivery of a new Beaufighter VI, so asked his CO if he could have a chat with Max Aitken and incorporate the new colour scheme on his aircraft. Permission was granted, the aircraft repainted in the new colour scheme, a new GM2 gunsight fitted, ready to take on the Hun on better terms. The opportunity for him and his RO* in X7773 came with their first sortie, so I will leave the story to John:

---

* Pilot Officer Griffiths.

We went off on a normal patrol, the moon was full . . . we were just mooning around on patrol when suddenly the controller said 'There is a bandit coming in from the east, he's probably at 20,000 ft' (We were at 18,000 ft). They gave me a vector to steer.

Suddenly my Nav/Rad shouted contact and gave me a vector. We were in business. After several changes of course we were behind him but overtaking fast (no visual yet). I daren't throttle back too quickly so I lowered the undercarriage and put on flap. That really pulled us up.

We were now synchronised in speed and the target would be about two hundred feet above me slightly to port. I increased speed a bit and suddenly got a visual. It was a Do.217 slightly to port and just above me. An ideal position. I turned on my GM2 gunsight and all the armament. He was going to get four 20 mm and six machine guns right up his jacksie. It wouldn't matter if he started to turn in wounded desperation, I had my new gunsight!

Well, life never turns out to be what you expect. I pulled across behind him at about 200 yards, and was about to pull up and open fire when the ventral gunner fired about fifty rounds of 13 mm armour-piercing incendiary. He hit us first time. The port engine burst into flames and a shell came through my windscreen, just left of the armoured bit of windshield, missed my head by inches and tore through the main electric cable and hydraulic pipe that ran down just behind the pilot on the left-hand side.

Now three things happened. We lost all our electrics, and the hydraulic oil spurted out all over me, the inside of the windscreen and the instrument panel. I was completely blind. Then my Nav/Rad called up and shouted 'Look out he's coming down on your tail' (a fully loaded bomber on its way to bomb Hull). Unable to see, no electrics or hydraulics (the guns probably wouldn't have fired) and my port engine blazing away. He who fights and runs away . . . I stuffed the nose down, pushed the fire extinguisher button for the port engine and we lost about 15,000 ft in a matter of seconds. I thought the Beau might fall apart.

At about 3000 ft I levelled out. The fire was out but the engine stopped. All the instruments were glittering with the oil. My Nav/Rad came forward and wiped the inside of the windscreen so that I could see out. We wiped some of the instruments. We were steering roughly 270 degrees heading for the English coast and were down to 2000 ft. I called for an emergency homing which I received.

The homing took them through the Hull balloon barrage . . . which they fortunately got through. The next emergency was a wheels-up, flapless landing at Church Fenton, making a long slow approach, missing the church near the end of the runway; the Beaufighter sliding along the ground, bumping and grinding to a halt . . . with John Wray's final remark: 'I hit my face on the gunsight but didn't do too much damage.'

By June 1942 six more squadrons of Fighter Command had been equipped with Beaufighters, these were 96, 125, 153, 256, 410 and 488. The Mk. VI was by this date the equipment of numerous squadrons and was receiving AI equipment up to Mk. VII standard. With this plus the improvement in operating, the radar chain and communications there was an increase in night interceptions and 'kills'; unfortunately, there were also more attacks by 'friendly' aircraft.

One of the kills recorded was by a 68 Squadron Beaufighter IF fitted with AI Mk. VII and flown by Squadron Leader V. Veseley (Czech), who had a fellow-countryman, Sergeant Necas as his RO. The action took place on the night of 23/24 July 1942; the pilot had followed a number of vectors given him by Control, gradually losing height to 3,000 ft, when the RO obtained contact of an aircraft slightly above at two mile range. Range was then closed until a visual was made at 300 yards . . . the enemy aircraft was an unusual visitor, being a He.177 that was jinking port and starboard and changing altitude continually. Due to cloud the chase continued and the range closed to 100 yards, when Veseley gave a full blast of gunfire from directly astern. The strikes on the He.177 blinded Veseley and return fire passed below the Beaufighter, so a break to port was made. When last seen the He.177 was diving vertically towards the sea from about 500 ft; it was considered that such a large aircraft diving from

this height could not recover and so was classed as destroyed.

Four days later another 68 Squadron Beaufighter IF, also fitted with AI Mk. VII, and flown by Pilot Officer Welch took off from Coltishall and was vectored onto an EA flying at 11,000 ft. The RO guided the pilot onto the contact, which turned out to be an intruder Me.110. The EA saw the Beaufighter and attempted to get on its tail, but after a number of turns made off at high speed. GCI then directed Welch onto another bandit, this was at 8000 ft and turned out to be a Do.217. A blast of cannon registered on the mainplane starboard centre-section, and the Do.217 turned on its back with a burning piece of the mainplane detaching itself as the aircraft went straight down.

Control next put Welch onto another contact, but after a long chase the AI radar was not working correctly, so a request was made to return to base. On the way back a contact was made at 5000 ft range and at the same height as the Beaufighter; so range was closed and the EA identified as a Do.217. The EA then opened fire and started to lose height and turn to starboard; the Beaufighter followed the Do.217 down and closed the range to 600 ft astern, and fired two bursts as he closed the range to 400 ft. The combat report stated: 'The second burst hit the Dornier in the centre of the fuselage and many strikes were observed, return fire ceased. After

Night-fighter IF X7583 of 68 Squadron prior to starting up, trolley acc' in position and crew hatches being closed.

losing visual, an explosion was seen by the RO as if an aircraft had crashed and was burning with the ammunition exploding'.

Not all were perfect interceptions and results, as Pilot Officer Salman and his RO Sergeant Perfect were to find on the night of 25 October. Flying a 141 Squadron Beaufighter I with AI Mk. VII, they took off from Ford and were vectored on to a Ju.88. There followed a series of chases, jinking, circling, climbing and diving, during which the EA opened fire. The Beau closed to 350 yards, then 300 yards and fired without any apparent effect, receiving back accurate return fire. During one last burst, having closed to 200 ft, return fire ceased and the EA pulled up steeply, did a stall turn to port with black smoke pouring from the port engine; contact was then lost when the Beaufighter dived after the EA. Returning to base the Beaufighter's control suddenly went slack, then whilst circling Ford the port engine caught fire; 25 degree of flap was selected, and as the undercarriage was down, this was re-selected up and a landing approach made. Before touch-down the starboard engine cut and the aircraft struck the ground short of the airfield, with both engines on fire; the crew escaped even though the aircraft was damaged Category B . . . it was a good landing, they walked away from it!

By now a number of squadrons were in the process of exchanging their Beaufighters for Mosquitoes, one of these was 307 Polish Squadron, who were by January 1943 flying both Beaufighter and Mosquito aircraft on night sorties. The last Beaufighter sortie by the squadron took place on 31 January flown by Flight Sergeant Wojczynski and Sergeant Sluszkiewicz on X8148. Taking off at 2010 hours the sortie was uneventful, no contact and no interception, with a landing back at base being made at 2200 hours.

During April 1943 a decision was made at Fighter Command to allow 'Ranger' long-range offensive patrols over *Luftwaffe* air bases, and for this 604 Squadron were moved to Scorton in Yorkshire, where they were re-equipped with ex-219 Squadron Beaufighter VIs fitted with AI Mk. VIII radar — the night fighters were to move into enemy territory.

In the meantime the night-fighter squadrons still had to continue the home defence of the United Kingdom, for raids were still providing 'trade'. Thus on the night of 23/24 May 1943, crews of 409 Squadron were sent on interception from their base at Acklington to intercept raids approaching the Tyne area. One of the crews was Squadron Leader G. Bower RCAF and Sergeant Beynon flying a Beaufighter VI equipped with AI Mk. IV.

Taking off at 0240 hours they were ordered by the Ouston Sector controller to 10,000 ft and given a vector to steer, at 9000 ft they got a head-on view of a twin-engined aircraft! When Bower asked GCI for instructions they were given a different vector to steer and a different target, almost immediately they obtained an AI contact and gave chase; the target was at 8000 ft and the Bandit dead ahead and slightly above at 700 yards. Gradually the range was closed and the target identified as a Dornier 217.

Closing in slowly the range decreased to 250 yards . . . then to 50 yards . . . a short burst of cannon fire and the Dornier began evasive action and returned the fire. A second burst of cannon fire and hits could be seen registering on the starboard wing and the return firing ceased. The third burst of cannon fire set the starboard engine on fire and what appeared to be flares were jettisoned. Almost immediately the evasive

manoeuvres ceased and the aircraft went into a gentle dive with the fire spreading, then at about 3000 ft the Dornier rolled over to starboard and dived vertically into the sea, leaving a large burning patch on the surface.

By this date a number of squadrons had been allocated for re-equipment with Mosquitoes, one of these was 488 Squadron RNZAF commanded by an 'ace' fighter pilot, Wing Commander J. Nesbitt-Dufort DSO. This squadron had been the last to convert on to Beaufighters, and was at the time of re-equipment flying the Mk. VIF version. It received its first Mosquito at the end of June, but would fly both types during the changeover period from Ayr on non-eventful patrols; night 'trade' by now being on the decline.

Some of the other units that were to equip with the Mosquito were 29 and 96 Squadrons, whilst 219 and 256 Squadrons had either already converted to Mosquitoes or been posted to the Middle East, as will be seen in the next section. Although most of the Beaufighters still flying had been equipped with AI Mk. IV, by now AI Mk. VII and VIII were re-equipping the Beaufighters and being fitted in the new production aircraft; the improvement in presentation was noticeable and the minimum range reduced.

By the summer of 1943 there were only six Beaufighter squadrons actively left in Fighter Command, these were 68, 125, 141, 406 RCAF, 409 RCAF and 604 Squadrons. As night 'trade' over the United Kingdom had fallen off, and some of the squadrons had aft scanning radar, 'Serrate', fitted, intruder patrols in the Bomber Command bomber stream were authorized. It was hoped that in this way a measure of protection could be offered to the bombers, by the interception of the German night fighters who were feeding into the bomber stream. The scheme was that even if the AI failed to pick up a night fighter, an enemy night fighter might be enticed onto the Beaufighter's tail in the mistaken belief that it was a bomber; the Beaufighter pilot would then wait until it was within range, then execute a smart 360-degree turn placing it, hopefully, on the EA's tail . . . thus the hunter becoming the hunted.

The first use of the Beaufighter, 'Serrate' and the tactic was carried out by six aircraft of 141 Squadron under Wing Commander 'Bob' Braham, on the night of 14/15 June 1943, when the main bomber stream was set for Oberhausen. The six Beaufighters flew into Coltishall to refuel, and set course from there for enemy territory; Wing Commander Braham on that night claiming one Me.110. Then on the night of the Peenemunde raid on 17/18 August a further success was claimed; five Me.110s of IV/NJG1 led by *Oberleutnant* Schnaufer intercepted on their radar what appeared to be stragglers from the bomber stream near the Dutch coast . . . within a few moments the trap was sprung and three Me.110s had been shot down . . . a further one crashed at base due to damage or weather. Over a few weeks 141 Squadron destroyed twenty-three German night fighters, Braham claiming nine with the help of his RO Flight Lieutenant Gregory.

This decoy scheme had originally been worked out by FIU at RAF Ford, where 'Monica' (active tail warning radar) had been developed under the code-name 'Whiting'. The 'Whiting' manoeuvre carried out by Braham* took its name from this development. The EA having been picked up on the tail-warning radar would be allowed to close its range to about 5000 yards (or 6000 yards if you were allergic to 20 mm cannon shells) before executing the 'Whiting' manoeuvre. A competent RO with 'Serrate', under good conditions, could pick up a 'Lichtenstein'-equipped EA on a reciprocal course at 10 miles.

Wing Commander Braham was striking again on the night of 17/18 August 1943, when with Flight Lieutenant H. Jacobs as his radar operator, he took off from Coltishall on an intruder patrol to Stade. At approximately 2300 hours the patrol was north of Schiermonikoog when they sighted a Me.110 flying east and jinking. Braham made a turn to follow the EA and closed to 300 yards range and opened fire with a two-second burst. Smoke came from the EA's port engine and then the EA dived to port . . . 'We gave him another two-second burst from 250 yards and he caught fire and dived into the sea, burning on the water'. Immediately afterwards another Me.110 was seen that had been tail-chasing the Beaufighter, so turning gently to starboard Braham drew the EA into his sights and gave the EA a one-second burst of cannon and machine-gun fire at 50 yards

---

* Wing Commander John 'Bob' Braham DSO, DFC, AFC.

'Ace' team of 141 Squadron, W/Cmdr Braham and R/O F/Lt Gregory, shown in a Mosquito.

*P/O P. R. Smith and P/O K. R. Lusty with their
Beaufighter R2093 of 25 Squadron, 1941.* (K. Lusty)

during a gentle turn. The EA appeared to blow
up, and a sharp pull-up and turn to port had to be
made to avoid ramming it. One man was seen to
bale out as the Me.110 dived vertically into the
sea in flames.

As has been said before, night fighting was
devious, scientific and calculating, and this
applied equally when the night fighter went
intruding; the crew used their Mk. I eyeball, their
AI and a cool approach to stalking their victims.
On 6/7 September 1943 the combat report of
Flying Officer White and Flying Officer M.
Allen hardly reflects this, but behind their sparse
report was hidden an intense and cool approach
with their Mk. I eyeballs fully tuned in. The 141
Beau, with the above crew, took off from Ford at
2045 hours for an intruder patrol to Juvincourt.
Their landfall was made at Ault and they flew
straight to the target area, although the ground
haze restricted the ground view. At 2210 hours
the crew saw four bright exhausts of an unseen
aircraft 5000 ft away flying east from Reims at
17,000 ft and turning to port. This was next
identified as a Ju.88 flying at a speed of about
160 mph and obviously not aware of the
Beaufighter's presence . . .

Throttling back and getting immediately
behind and on the same level we gave a
two-second burst of cannon and machine-
gun from 600 ft, closing gently to 550 ft.
There were many strikes on the fuselage
and the starboard engine, which burst into
flames. Burning petrol came streaming
back and small burning pieces came away
in showers. Streams of glycol covered our
windscreen. We gave EA another two-
second burst with all guns from the same
range and there seemed to be strikes all
over it. EA climbed steeply, straightened
out and then dived steeply in flames. We
watched it strike the ground, where it could
be seen through the haze burning fiercely.

An interception with a difference took place
on 6 September, with 406 RCAF Squadron based
at Valley (Anglesey). At 0840 hours Squadron
Leader Williams and Pilot Officer K. Lusty were
scrambled in 'P' to intercept a 'plot' which was
behaving strangely. When a visual was obtained
it turned out to be a Flying Fortress, over the
Irish Sea and flying north. Ken Lusty recounts:

We flew alongside, but could not see any
member of crew. Normally the gunners
would follow us (with their guns) and the
rest of the crew would make rude gestures
from the flight deck. Not a sign of anyone!!

We flew to port and starboard, behind, below, above and in front of the aircraft. We 'waggled' our wings but there was no sign that we had been seen or in fact anyone was on board. No escape hatches were open and no sign of any damage to the plane at all. It was a very peculiar feeling being up there with what seemed to be an empty plane. Eventually the Isle of Man appeared below and Dave Williams decided that if anyone was on board, they must now know where they were.

The Beaufighter returned to base and the crew never knew the end of the story — perhaps it was never known — but later on, when they read of the *Luftwaffe* repairing Allied aircraft and flying them on all manner of missions (KG200), they did wonder if this aircraft might have been one flown by a German crew on surveillance or photography.

September also found Braham completing his tour, and on the night of the 29/30th he chalked up his last 'kill' on Beaufighters, shooting down a Me.110 flown by the *Luftwaffe* night-fighter ace *Hauptmann* August Geigt of NJG1. It also saw 141 Squadron start to convert from the Beaufighter to the Mosquito. Another squadron that was in the process of conversion was 515 Squadron under Squadron Leader S. Thomas. It was considered by now that intruder night-fighter sorties in Bomber Command raids could be more effective with the Mosquito, and that it was also a more manoeuvrable aircraft with longer range. This latter point could not be contested, but although the newer Mosquito would have a higher maximum speed and ceiling than the Beaufighter, the latter had a much better rate of climb, and in strength it was unbeatable.

On the night of 3/4 October a Beaufighter VI of 68 Squadron put paid to a Me.410. The crew of the Beau were Flight Lieutenant P. Allen and Flight Lieutenant N. Josling and the aircraft was equipped with AI Mk. VIII radar. That night whilst flying at 10,000 ft the pilot saw an aircraft coned in the searchlights and recognized it as a Me.410; the Beaufighter itself was coned by searchlights, and only managed to extricate itself by firing the colours of the day, signalling and evasion. The search continued and contact was made with the EA at 2–3 miles at speeds of 290–330 mph, with the chase lasting for just over fifteen minutes, over the North Sea on a course

SE of Yarmouth. The Beaufighter then closed the range to 2000 ft and a visual was obtained, range was then decreased to 1500 ft and a seven-second burst of cannon fired. Strikes were seen on the starboard wing, which was immediately followed by an explosion in the wing. The EA then dived trailing black smoke and exploded on contact with the sea.

As autumn approached a number of squadrons had converted from the Beaufighter to the Mosquito; this change had been forewarned early on in the year, when C.L. Courtney* in a letter to Air Vice-Marshal Sir W. Freeman, dated the 10 February 1943 wrote: '. . . decided to accept the Mosquito as the night fighter for Fighter Command, and night fighter and long-range fighter for overseas. This is admittedly a slight gamble, but I understand that your technical experts have no reason to believe that the Mosquito will fall to pieces in hot climates'. The latter part was to prove not strictly true as regards the hot humid climate in the Far East, as can be seen in R&M2600.

As a result of this decision the Beaufighter was ultimately removed from Europe as a night fighter, so that as D-Day approached the Beaufighters of Fighter Command were supplanted by Mosquitoes, and the last six squadrons were re-equipped. 141 Squadron had been the first to go, with the last squadron, 406 RCAF Squadron, receiving its Mosquitoes in April 1944.

406 Squadron had by December 1943 been transferred to Exeter, still commanded by Wing Commander R. Fumerton DFC, and were operating Beau VIs fitted with AI Mk. VIII. From Exeter ASR patrols as well as protection and interception patrols were flown. December went out on a happy note, when an ASR patrol flown by Warrant Officer McConnell and Flying Officer M. Kazakoff located the dinghy of a ditched Flying Fortress, circling it while transmitting position to base. They were relieved by another of the squadron's Beaufighters, who watched an ASR boat reach the survivors.

Nearing the end of the squadron's use of the Beaufighter, came the destruction of a rare target. On the night of 19/20 March 1944, KW101 flown by Squadron Leader Williams and

---

* Retired as Air Chief Marshal Sir C.L. Courtney CBE, KCB, DSO (AMSO).

Flying Officer Kirkpatrick, were vectored south over the Channel after a 'bogey'. Contact was made approximately twenty miles off Guernsey, then closing in fast they obtained a visual of EA at 600 ft, this was identified as a He.177. Closing the range until quite close, Williams opened fire, the two engines exploded and the aircraft went into a vertical dive.

Returning to 1943, during the summer of that year, four squadrons of the USAAF, 414th, 415th, 416th and 417th, had commenced night-fighter training on Beaufighters with the RAF at Acklington, Ayr, Cranfield, Honiley and Scorton. With relatively 'green' pilots and a powerful aircraft with swinging tendencies, there was some charging across the grass and some bent airframes. Just the same, four competent USAAF squadrons were operational on their Beaufighters by September, when they departed to the Middle East; there they formed an effective night-fighter force of the 12th Air Force in North Africa. They would continue their nocturnal activities as the Allied Armies advanced through Sicily to Italy. The 415th Squadron, still operating their Beaufighters, moved to France in September 1944, and then operated under 2nd TAF. These were standard Mk. VIF aircraft in standard RAF camouflage — though they may have been equipped with cigar-lighters by then!

By then the night-fighter Beaus had moved on, fresh fields to conquer, bequeathing to the newer Mosquito and its crews a wealth of experience

and technique gained through blood and sweat — this was the Royal Air Force's first *real* night fighter . . . the BEAU.

Coastal search and strike

The use of the Beaufighter by Coastal Command as the Battle of Atlantic developed became an operational necessity, because the aircraft was the only viable type in the strike-fighter role. On 7 December 1940 the Air Ministry notified Headquarters 43 Group that the issue of Beaufighters to 252 Squadron was to take top priority. The beauty of the Beaufighter for Coastal Command was that it was especially suitable because, by the addition of special equipment, very little modification and no major alteration, the design was admirable for the role of long-range escort fighter. In regard to the introduction of the Coastal Command version, steps were taken to avoid disturbing the flow of Fighter Command aircraft. Changes were not introduced concurrently into the production line, but treated as a series of modifications which were introduced as the opportunity occurred.

To recap on Coastal Command's operational capability on the commencement of hostilities in 1939; it should be accepted that for a seaborne nation dependent on control of the seas for the majority of our imports (not just food, but fuel and oil for our Fighter and Bomber Commands), its control of the air above those seas was precarious — and this implies no criticism of Coastal Command's air and groundcrews. Its coastal reconnaissance was operated by the Avro Anson, gradually to be supplemented and then superseded by the Lockheed Hudson which

Mk VIF Beaufighter of the USAAF serving in the Middle East.

Mk IC R2153 of 252 Squadron at RAF Aldergrove 1940 awaiting take-off clearance. (Paddy Porter)

extended the range into the Atlantic and Northern seas. By 1940 the only type of strike aircraft that the Command possessed was the Bristol Beaufort; this, though modern, could not operate in an enemy-controlled environment without fighter escort, and in its bombing or torpedo-carrying role was dependent on darkness or cloud cover for any immunity from the enemy. Handley Page Hampdens would be later converted to the torpedo bomber role, but this was also vulnerable; as were the Bomber Command Blenheims transferred for North Sea sweeps. So the development of the Beaufighter, with its strength and fire-power, was an absolute necessity for the equipment of strike squadrons, to allow offensive strikes across the North Sea.

The Coastal Command version of the Beaufighter originated with a request from Coastal Command for a long-range fighter, to take the fight to the enemy, who were attacking British shipping on the high seas as well as in the North Sea. So Bristol's modified Mk. 1 R2152, removing the wing guns and installing fifty-gallon fuel tanks instead, fitting a navigation table for the Nav/RO, and replacing the camera gun with a DF loop; as modified the aircraft was listed as a Mk. IC. A further conversion was made on R2269 as a trial installation, then followed by completion of ninety-seven aircraft for this role; the first one, T3228, joined 252 Squadron on 8 March. The first Mk. ICs were minus the extra fuel tanks, as suitable ones were to be produced by a subcontractor as soon as possible; but as an interim measure, fifty-gallon

Wellington fuel tanks were installed temporarily on the floor between the cannon bays.

As stated previously, 252 Squadron was the first to receive the Beaufighter IC; at the time they were based at Chivenor working up to operational fitness. The unit later moved to Aldergrove to commence its operational debut, flying the first operational sortie from there on 6 April 1941. Then on 25 April orders were received to dispatch fifteen of their Beaufighters to Malta, complete with air- and ground-crews. This left the squadron in a rather parlous state regarding aircraft and crews, but they still carried on flying operational sorties until June, when 143 Squadron took over the remains of 252 Squadron — and so 252 Squadron disappeared from the Coastal Command scene.

By the start of 1942 there were still only three squadrons of the Beaufighter in Coastal Command, these were 235, 236 and 248 Squadrons, and they were not all wholely equipped with Beaufighters. 235 Squadron had been transferred to Coastal Command in early 1940, but did not receive Beaufighters until December 1940, and continued to fly both Blenheims and Beaufighters well in to 1941.

With the secret operation on St Nazaire to block the dock gates, code-named 'Operation Chariot', 236 Squadron supplied Beaufighters to help provide air cover. The aircraft operated from Predannack and flew to the French coast singly; during these escort sorties T4915 sighted

Mk X of 236 Squadron armed with rocket projectiles out on a sweep.

four destroyers, and after giving recognition signals was still fired on; as it was normal procedure to be fired at by the Royal Navy, the reception left some doubt about the identity of the destroyers. Later on the one engine started failing, so the aircraft returned to Predannack. T4796 arrived off Ushant but could find no convoy, it was attacked by a single-engined enemy fighter without results on either side.

Next in was Wing Commander Glover flying T4755; during his sortie he sighted a He.115 flying at sea-level, which was probably on reconnaissance searching for the 'Chariot' convoy returning. Glover gave chase but the EA climbed fast for cloud cover, and could no longer be located. Squadron Leader Pike in T4938 was next on the escort duty, and upon sighting the convoy also sighted a He.115, which was apparently 'shadowing' the vessels. Pike closed on the EA and fired, but the EA went into cloud cover towards St Nazaire. Escorting the vessels Pike again sighted a He.115, which he managed to attack; hits were seen on the EA, but return fire hit the port side of the cockpit area and severed the hydraulic system pipeline. Combat was broken off and a return made, with a force-landing at St Eval due to failure of hydraulics.

T4795 crewed by Sergeants Taylor and Parfitt took over escort of the convoy returning from St Nazaire, but failed to return. Naval authorities reported the Beaufighter running into a force of

Ju.88s and engaging one. The EA blew up in mid-air and hit the Beau, which then crashed in the sea.

After this up to May the Squadron were engaged on recces of the Dutch coast for enemy convoys, which although fairly monotonous — apart from occasional flak bursts to 'shoo' the Beau off — a target did sometimes show up. On 10 May Pilot Officer Lee and Flight Sergeant Taylor in T4917 were returning to base after a recce of the Dutch coast, flying at 50 ft above sea-level, when they sighted a Do.217 also at sea-level about two miles distant. Lee gave chase and closed on the EA, and upon opening fire and missing, the Dornier carried out evasive action. So making another approach from slightly below, Lee fired a two-second burst and saw hits on the starboard wing root, followed by pieces flying off. The EA then began to fly erratically, but T4917's port engine started to fail, so combat was broken off and a return to base made.

As a break from recces of the Dutch coast, two 236 Squadron aircraft on 30 May went on an intruder sortie from Wattisham. X8060 flown by Squadron Leader Jay and Sergeant Kent took-off at 2348 hours and flew inland over Holland, circuited Leeuwarden airfield but saw no aircraft. They were then engaged by light flak and searchlights, so carried out evasive action and returned unsuccessful. X8036 flown by Wing Commander Wood and Flight Sergeant Taylor took-off at 2355 hours on a heading for Texel airfield for a possible intruder combat. However, there were no interceptions or incidents, but an early return to base was made as the starboard engine began overheating.

236 Squadron were again in the news in June, when on 12 June Flight Lieutenant A. Gatward* and his RO Sergeant G. Fern flying T4800, took off from Thorney Island and flew at deck-level in daylight to Paris. There they thumbed their noses at the *Luftwaffe*, by dropping a French *Tricolore* on the Arc de Triomphe and then attacked with cannon the German *Kriegsmarine* Headquarters in the Place de la Concorde.

June also saw the conversion of 254 Squadron into Beaufighters, the unit being based at Dyce under the command of Wing Commander R.H. McConnell DFC. This unit had been flying Blenheims, and would fly a mixture of Blenheims and Beaufighters before becoming operational in November 1942. This long period of re-equipment was symptomatic of the time, with priority for the Beaufighter to be in Fighter Command or overseas; 236 Squadron for instance, in February 1942, had been divested of its experienced crews, who were posted to the Middle East; new crews and Beaufighters were then assembled for training in the anti-shipping role.

Meanwhile 143 Squadron had been employed carrying out an OTU function and training crews for overseas, and gave up its Beaufighters to 235 Squadron. May 1942 saw the squadron starting to receive new Beaufighters, which by August were declared operational; the squadron was then commanded by Wing Commander E. Thornewill. In that short space of time the

squadron had been stationed at Limavady and Thorney Island, and was now stationed at North Coates; its main task at first was 'Rover' sorties along the Dutch coast and on recce missions.

At the start of 1942, with Bristol's proposal for the Beaufighter torpedo-bomber before them, the Air Ministry were considering the specification and operation, with the Admiralty sitting in. This was followed by discussions between the Admiralty, Air Ministry and Ministry of Aircraft Production about the strike squadron requirements, replacement of the Beaufort torpedo-bomber and the type of replacement aircraft. It was not until July before a meeting of MAP and Air Ministry finalized the decision, when it was stated that the Admiralty accepted the principle that the day-torpedo-bomber should be in the fighter class, and it was agreed that a conversion of the Beaufighter appeared to be best. It was considered that the Mosquito could take over the night-fighter role in place of the Beaufighter, and the torpedo-bomber Beaufighter would take the place of the Beaufort.

235 Squadron were at this time acting in the long-range fighter role, and finding their prey over the seas bordering the French, Dutch and Belgian coasts. On 26 August, two of their Beaus

---

* He was eventually to become Wing Commander and CO of 404 Squadron, and retired as Group Captain DSO, DFC.

Mk IC T4800 of 236 Squadron flown by F/Lt A. Gatward on 12 June, 1942, low level over Paris to drop Tricolore.

were out on patrol; these were 'M' flown by Sergeants Woodcock and Colman, and 'Y' flown by Pilot Officer Schaefer and Sergeant Lawton. They sighted a Ju.88, but 'Y' lost contact in the cloud. 'M' located the EA and closed to 100 yards and opened fire; strikes were seen on the Ju.88 fuselage, but the EA carried out evasive tactics and disappeared into the clouds. In searching around 'M' then located a FW.200 at 150 ft above the sea south-west of Brest, so closing up on the EA to 300 yards Woodcock made a stern attack and saw strikes on the EA. The FW200 immediately dived away and was last seen at sea-level.

Acceptance of the Beaufighter as a strike aircraft was made by the Air Ministry in September 1942, and finalization of an agreement was made to adopt the aircraft as Coastal Command's strike aircraft in the day torpedo-bomber. The intention was to form three Strike Wings, with each Wing composed of a flak-suppression squadron, a fighter-bomber squadron and a TorBeau* squadron. In the

original planning these Wings were to be based at North Coates, Leuchars and Wick, but due to priorities in other parts of the world's war zones, the actual creation of the Wings and their operations was delayed.

Formation of the first Wing took place at North Coates during November 1942, and the units nominated were 143, 236 and 254 Squadrons. For shipping strikes it was intended that fighter cover would initially be provided by Spitfires for short-range operations, and Mustangs when they became available. The Wing's first operation was carried out by 236 and 254 Squadrons led by Wing Commander Fraser† of 236 Squadron. The strike was made against a convoy off the Hague on 20 November 1942 and was unfortunately not an outstanding success. The attack was not very well co-ordinated, contact with the fighter escort was not made, and resulted in three Beaufighters being lost and a number of others damaged . . . for the sinking of a small tug!

---

* This was not originally an official designation, but a nickname bestowed by the squadron's personnel, and was generally adopted.

† He failed to return from this operation.

The failure of the strike resulted in the withdrawal of the Wing from operations and the commencement of an intensive training programme, as well as a report to the Air Ministry. The TorBeau and its deployment coincided with the enemy increasing the number of flak-ships per merchant ship, as well as the increase of flak defences on the ships — for instance, an armed trawler would have at least one 88 mm gun with five to seven weapons of 20–37 mm. So the training was concerned with the suppression of the flak defences, getting the TorBeaus into the launching position and then exiting as quick as possible. It was also determined that the flak-suppression aircraft were to have their attack synchronized with the torpedo attack, and in the interest of manoeuvring and collision prevention, it was inadvisable for more than four torpedo aircraft to concentrate on one merchant vessel. Once exiting, it was out fast and low between the enemy ships, so that the enemy flak gunners would be restricted in fear of hitting their own ships.

By January 1943 there were in Coastal Command nine units equipped or equipping with Beaufighters; these were 143, 144, 235, 236, 248, 252, 254, 272 and 404 Squadrons. The last unit had commenced equipping with Beaufighters in the autumn of 1942 in the role of a long-range fighter unit. 404 Squadron was in early 1943 based at Chivenor, but at the start of April it moved to Wick in Scotland to provide anti-shipping escort for TorBeaus.

The 1 March found Sergeants Vanderwater and McLachlan of 235 Squadron taking off on patrol in weather that was three-fifths cloud and visibility about 5 miles. About two hours after take-off a Ju.88 was seen circling at 1000 ft, and made off as the Beau approached. A 600-ton vessel named *Tromosunde Norge* was then found flying three flags with three lifeboats full of people wearing yellow life-jackets alongside. The Beau circled and signalled the direction to steer. A Ju.88 that tried to bomb the ship was attacked by the Beau and the EA dived to sea-level with its port wing on fire; a second attack by the Beau from the beam, and the Ju.88 somersaulted and disintegrated into the sea.

On 22 February a meeting was held at Coastal Command HQ, the point of discussion being the operation of the 16 Group Strike Fighter Wing and its co-operation with Fighter Command aircraft. It was decided that two single-seat fighter squadrons (twenty-four aircraft) should escort each Strike Wing, one as a close escort and one as top cover. If the cloud base was below 2000 ft in the target area, then not more than two escort fighter squadrons were to be employed, and these would fly as close escort. On all operations the decision was made that the escort and strike leaders should be in VHF communication with one another. One further decision was that the 'Jim Crow' aircraft (code-named LAGOON) was to land at the Strike Wing airfield if it had sighted any enemy shipping. In the opinion of all present all fighter and RP fighter aircraft should be directed at the escort vessels, to leave the task of sinking the target vessel(s) to the TorBeaus.

On the basis of this and after their period of re-forming and retraining, the North Coates Beaufighter Wing carried out its first attack on enemy shipping on 18 April 1943. This strike was led by Wing Commander N. Wheeler*, and consisted of twenty-one Beaufighters from 143, 236 and 254 Squadrons. The fighter escort was provided by 118, 167 and 613 Squadrons flying Spitfires and Mustangs. The strike set off and found the convoy off Texel, the attack went in as planned, resulting in the sinking of a large vessel, the *Hoegh Carrier*, as well as damaging other vessels. After this strike further attacks were put in, and from intelligence sources it was ascertained that the strikes were not only causing a lot of damage to the enemy, but forcing him to increase the number of escort vessels.

At this time, both the Command and Slessor, their AOC-in-C, were getting frustrated with the continual drain of newly-trained Beaufighter aircrew and their aircraft to the Middle East theatre of operations. With Harris of Bomber Command forever requiring more and more aircraft, and Fighter Command having priority on fighters for the defence of the UK base, Coastal Command was becoming — or was already — the Cinderella of the service when it came to supplies. Then on 17 April the Planning Division of the Naval Staff at the Admiralty wrote to the Air Ministry asking for an extra torpedo-bomber squadron for the Middle East for

---

* Later, Air Chief Marshal Sir Neil Wheeler GCB, CBE, DSO, DFC, AFC.

'Operation Husky', suggested date of departure 20 June, but the Squadron not to be withdrawn from its operations until the last minute!

The withdrawal of one TorBeau squadron from Coastal Command operations meant that there would then only be one such squadron left in the UK. Nevertheless, it was on orders to comply with this request and 144 Squadron was nominated. The squadron was withdrawn from the line on 3 May, their Servicing Echelon and supplies prepared for temporary detachment for twenty days and sixteen Beaufighter X aircraft adapted for the tropics nominated for the flight to North-West Africa. These left from Portreath and transited through Gibraltar to Port Lyautey; by the time that they got to their base at Protville there were only twelve Beaufighters serviceable; four had been shot down *en route* or had broken sternframes.

Another shipping strike was made off Den Helder by the North Coates Wing on 18 July, but was considered by Coastal Command HQ as practically abortive — doubts were being voiced at conferences as to the effectiveness of the attacks, there being clear evidence in statistics

Beaufighters of 254 Squadron cannoning and rocketing armed escort vessels during a shipping strike in 1944. (Paddy Porter)

that Bomber Command mine-laying was claiming far more enemy ships sunk than by Coastal Command strike aircraft. What was not so apparent from such statistics was the energy required by the enemy to counter the strike aircraft, and the increasing numbers of defence vessels as escorts and the diversion of aircraft for their defence. This and the 'Channel Stop' attacks by Fighter Command aircraft had reduced the enemy shipping along the enemy-held coasts to minimal proportions, and where previously, Swedish vessels had freely plied their trade to German ports, crews and owners were now reluctant to hazard themselves and their ships.

Prior to these conferences 235 Squadron was employed on long-range fighter activities, in some cases providing escort for Hampdens of 455 Squadron. In one case on 12 April having to defend the Hampdens from FW.190s. This occurred when on a strike in Norwegian waters, rain clouds forced the formation to climb to 4000 ft, and as they made landfall they could see the enemy fighters taking off from Sola airfield. The EA dived from astern of the Hampdens and zoomed up to make further attacks, and the Beaus tried to place themselves between the EA and the Hampdens and made successive interceptions, but 'Z', flown by Sergeants Elliott

and Pinkney with smoke coming from its port engine, went in to the sea. As the RAF formation pulled away one Me.109 and two FW.190s were claimed as having been put down, with another Fw.190 seen heading for the coast losing height.

The same squadron on 24 April had a clash with a U-boat and flak ship whilst again escorting 455 Squadron Hampdens. The flak ship opened fire on the Hampdens, and then heavy flak was sent up by shore batteries. So whilst an escort was maintained on the Hampdens, two of the Beaus went down and swept both the flak ship and the U-boat it was escorting; the U-boat's gun crews were seen to disappear under the firing from the Beau's guns and smoke and flames came from the flak ship's superstructure. Meanwhile the Hampdens had carried out evasive action, and as the formation pulled away the U-boat had submerged.

On 1 May 1943 the North Coates Wing were to suffer heavy casualties, when thirty-one Beaufighters were dispatched to attack the German cruiser *Nurnberg*, which had been reported south-east of Norway with a retinue of three heavy destroyers. 236 Squadron sent twelve aircraft under Squadron Leader G. Denholm, eleven TorBeaus of 254 Squadron, escorted by eight fighter Beaus of 143 Squadron. Heading for the Norwegian coast the Lister light

was identified, then a patrol closed the shore in the vicinity of Stavanger but no enemy ships were seen. This was immediately followed by the Beaufighter formation being attacked by a force of FW.190s, and heavy flak spewing up from the shore batteries. The Beaus of 143 Squadron put up a gallant defence of their charges, but were hardly a match for the nimble single-seat fighters. A signal was given by the formation leader for the Beaufighters to 'break' and return individually to base.

From the *mêlée* 143 Squadron lost JL943 flown by Warrant Officer V. Bain and Sergeant Room, JM108 flown by Flight Sergeant Foster and Sergeant Curnuck, and Flight Sergeant Baker the Nav/R of JL945 was killed. The strike resulted in three other Beaus failing to return without tangible results.

In spite of these losses the Wing carried on its operations, and on 13 June twenty-nine Beaus of the three squadrons were out on a convoy attack. This was located a few miles south of Den Helder, with the main target the *Stadt Emden* of 5200 tons amongst other vessels and mine-sweepers. Ten Beaus of 254 Squadron attacked the *Stadt Emden* in pairs, hitting and setting it on

254 Squadron Beau's break away after the strike on shipping. (Paddy Porter)

fire, with explosions seen on other merchant vessels and at least four mine-sweepers damaged. All the Beaufighters returned safely and photographic interpretation indicated that the *Stadt Emden* and another MV had been hit with torpedoes.

Until July 1943 the code-name for shipping recces was LAGOON, this was now changed to SHIPPING JIM CROW, with all operations laid on by the 12 Group controller at the request of the 16 Group controller. Then in August came a series of conferences to discuss the operation of the anti-shipping Wing. The North Coates Wing consisted of three Beaufighter squadrons, one of TorBeaus and two of anti-flak escort Beaufighters, the whole totalling sixty first-line aircraft and seventy-two aircrew. From the time that the Wing commenced operations up till the end of April 1943 there had been fifty-six enemy convoys run between the Hook and the Elbe, of these, fifty-five had been sighted but only nine attacks had been flown.

Some of this was due to adverse weather conditions, incorrect sightings etc., but the results to the Air Staff appeared to be poor returns for the involvement of three squadrons. This resulted in a further conference being called on 20 August to discuss the anti-shipping requirements in the North Sea and the Channel, the meeting being held under the chairmanship of the DCAS. It was now clearly stated that the locking up of the North Coates Wing of three Beaufighter squadrons, with the primary role of striking against enemy convoys, and the requirement for considerable numbers of fighter aircraft had produced low returns in ships sunk. So it was agreed that Coastal Command's torpedo-bomber squadrons should be reduced from five to four.

The following day 16 Group were discussing the use of the North Coates Wing and its operational formation. It had been found desirable that six TorBeaus should be employed against each motor vessel (which was different from the plan at the end of 1942), and against each escort vessel three cannon- or RP-equipped fighters — which meant that the North Coates Wing would on its own, only be able to take on a force of one motor vessel and four escort vessels twice a week!

Needless to say, whilst all the conferences were going on, the ground- and aircrews of Coastal Command kept up their work, re-equipping or receiving new equipment, seeking and striking where possible at any point where the enemy had his ships. Around this period for instance, both 143 and 236 Squadrons were having their Beaufighters modified to accept a new weapon, the three-inch rocket projectile. This would prove the most devastating addition to the Beaufighter's armoury, both in the anti-flak suppression role and ship sinking, as would be so effectively demonstrated later.

235 Squadron operating from Chivenor had two Beaufighters take-off on an offensive sweep, which just over two hours later sighted a large column of water as if bombs had been jettisoned, and also a Ju.88. Upon investigation an RAF-type dinghy was found with four crew, so a smoke float was dropped and a message sent to base. A stern attack was then made on the Ju.88, then a beam attack from 150 yards. From the two attacks strikes were seen on the starboard engine, fuselage and tailplane; the EA dived, levelled out as if to ditch, then the one Beau dived from 400 ft and sent in a burst from the beam and the Ju.88 went into the sea. Momentarily three survivors appeared, but then disappeared as the wreckage went under.

Coastal Command's task in the first three years of the war had mainly been the provision of escorts for surface vessels and convoys, with strikes at commerce raiders and other German shipping.* With the introduction of the Beaufighter there was now the possibility of offensive strikes with an aircraft capable of looking after itself. Then in 1943, with Coastal Command's anti-U-boat and anti-shipping heavy aircraft, such as the Halifax and Liberator, denying German shipping the right of passage, and also taking a roll of U-boats in the Bay of Biscay, the U-boat Command requested from the *Fliegerführer Atlantik* air protection for the U-boats. This took the form of Ju.88 fighters being drafted to the area, and resulted in the loss of British aircraft. In retaliation against this, Coastal Command sent in Beaufighters to carry out offensive sweeps of the area, followed by successful encounters with both Ju.88s and FW.190s.

---

* Blockade runners were not always easy targets, even on their own; in 1942 three Lancasters on loan to Coastal Command were shot down attacking one.

A typical example of such an offensive strike took place on 10 March 1943, when four Beaus of 248 Squadron led by Flight Lieutenant A.R. de l'Innis took off on patrol from Predannack. Their first sighting was a Ju.88, which was attacked, bursting into flames after Lieutenant J. Maurice-Guedj, flying 'W', closed in from astern to 150 yards and gave the EA a two-second burst. The EA straightened up out of its action and glided into the sea and crashed. 'W' however had received damage and crash-landed at base.

The effectiveness of the rocket projectiles was to be proved on 1 June, when T5258 of 236 Squadron, flown by Flying Officer M. Bateman and Flight Sergeant Easterbrook took off from Predannack for a sortie over the Bay. Just after 1100 hours the wake of a U-boat was sighted from 3500 ft at a distance of about ten miles, preparations for an attack was made and the distance closed, when it was identified as a 517-ton type U-boat. The U-boat started to submerge but Bateman was already diving to the attack from the port beam, releasing four 25 lb RPs at the enemy craft and observing the explosions. Circling round after the attack there was no sign of the U-boat, but a large green patch lay on the surface, followed shortly afterwards by another green patch and debris coming to the surface a bit farther away. U-418 was sunk approximately 130 miles west of the Brittany coast.

A further sortie by 236 Squadron two days later was not quite so successful, although it did prove the toughness of the Beaufighter. JL819 flown by Flight Lieutenant H. Sharon and Pilot Officer I.S. Walter took off from Predannack on an anti-U-boat patrol. About two hours later, whilst flying at 3500 ft, they were attacked by eight Ju.88s out of the sun. Three coming in on either beam and one from each quarter, firing at about 1500 yards. Sharon dived the Beau down to 1500 ft, but Walter had been hit by the first burst and couldn't use his gun. The Junkers then formed up and began to dive in pairs from each side, attempting to rake the Beaufighter as Sharon whipped the Beau about in evasive action above the sea.

Bullets and cannon shells raked around the Beau and Sharon felt a bullet pass his head, the Royal Navy observer behind the pilot was hit, receiving fatal head injuries. The Junkers dived at the Beau, firing and re-forming for another attack, but still Sharon weaved the Beau, twisting

and turning, and in one case making a pass at a Ju.88 that attempted a head-on attack. The EA attempted to make the Beau turn from its course, Sharon having now settled on a northern course for base, but still he weaved, evaded, still the Ju.88s came in on hot pursuit, but the two types of aircraft were evenly matched for speed, so that in the end the EA called it off and headed home; whilst JL819 with hydraulics, undercarriage and flaps not working, made a belly landing at Predannack.

August 1943 saw the return of 144 Squadron from the Middle East back into Coastal Command, first based at Tain and commanded by Wing Commander D. Lumsden DFC. Arriving back in the UK during the pros and cons discussions on the Strike Wings, the squadron was based at Wick, and provided escort to other aircraft on anti-shipping strikes off Norway. These were usually 'Rover' patrols off the Norwegian coastline in conjunction with 404, 455 and 489 Squadrons.

On 27 December 1943, such a 'Rover' towards the Norwegian coast sighted an enemy convoy of one 6000-ton merchant vessel, four small coasters, one 'M' class mine-sweeper and two escort vessels on a southerly course. During the attack hits were registered on the merchant vessel and strikes made on the mine-sweeper, but one Beau could not release its torpedo as the air-tail was damaged by flak. 'M' of 144 Squadron, crewed by Flying Officer Harrison and Sergeant Wilkinson, attacked a Bv.138 flying-boat escorting the convoy. Closing from the port beam, Harrison opened fire at 300 yards, closing to 150 yards. The EA caught fire and dived to the sea, levelled out and cartwheeled onto its back, still on fire.

As opposed to Bomber Command and Fighter Command, Coastal Command's operations were carried out unpublicized, yet amongst the squadron OR books and combat reports of Coastal Command can be found countless acts of heroism performed by crews during their anti-shipping strikes, that rival those of any other Command. None more so, than those of the Beaufort and Beaufighter squadrons; unselfishness in devotion to duty and in support of one another went unheralded, such as occurred on 26 January 1944.

Six Beaus of 144 Squadron and two Beaus of 404 Squadron took off to escort four RP Beaus of 404 Squadron; their target was a convoy of three merchant vessels. The enemy convoy was sailing with an escort vessel forward and astern and another seaward, travelling along the enemy coast, with one 'M' class mine-sweeper ahead. 144 Squadron made their attack on the seaward escort vessel and mine-sweeper, and strikes were made on both vessels. The mine-sweeper ceased firing and appeared to be on fire. 404 Squadron's anti-flak Beaus put in their attacks on the rear escort vessel whilst the RP Beaus attacked the merchant vessels. Soon the forward merchant vessel was on fire and slowing down with the other one damaged, and an escort vessel stationary in the sea.

As the Beaufighter formation broke off their attack, three Me.109s put in an appearance and attacked NE318 flown by Pilot Officer J. Dixon and Sergeant E. Pearce, which had just completed its attack on the leading merchant vessel; NE318 was crippled and crashed into the sea in flames. Next to be attacked was 'M' of 144 Squadron flown by Flying Officers B. Sanson and R. Sewell; Sanson tried to make cloud cover, but as the EA came up astern and fired, he corkscrewed the Beau to sea-level and began evasive manoeuvres, during which the EA overshot, when Sanson took a quick burst at him.

Seeing this action, NE339 of 404 Squadron flown by Flying Officers S. Shulemson and P. Bassett, turned back and engaged the Me.109, allowing 'M' to break clear as it had received numerous cannon strikes from the EA. With Bassett giving a running commentary on the EA's position and also using his Browning machine-gun, Shulemson was able to carry out violent evasive action with NE339, as well as take an odd burst at the Me.109, eventually making rain cloud cover and escaping. Both 144 Squadron's OR Book and Flying Officer Sanson's combat report make the point that 'M' would probably have been shot down had Shulemson not turned back to attack the Me.109. Shulemson's combat report was annotated by Intelligence that over the past six months of operations, this officer had shot down a Bv.138 and damaged a Ju.188.

144 Squadron during another 'Rover' off the Norwegian coast on 30 March was to fare better, although 404 Squadron was to lose two of their Beaus. Four TorBeaus of the squadron with an escort of three more and nine of 404 Squadron

took off from Wick in the afternoon. At 1825 hours an enemy convoy of one large merchant vessel of about 14,000 tons and an escort of one destroyer, one 'M' class mine-sweeper and a smaller escort vessel was sighted. The attack went in at 150 feet with the anti-flak Beaus firing off their cannon and RPs, and LZ538 detailed off to concentrate on the destroyer. Not only did the ships put up an intense barrage, but the shore batteries joined in as well; then with the approach of EA, LZ218 and NE227 were detached to give cover, resulting in a damaged Me.110. As the Beaus left the scene, the destroyer was smoking badly and the merchant vessel had taken two torpedo hits.

John G. Lingard* was widely experienced on the Beaufighter, and flew with Coastal Command, in the Middle East and Far East. Writing about the Beaufighter operations with Coastal Command he said:

> Strikes from UK airfields were flown at 500 feet, decreasing to 200 feet to avoid detection by enemy radar. Altitude was increased in carrying out the attacks on convoys.
>
> 143 Squadron carried out dive bombing from 2000 feet against Channel ports with two 500 lb plus two 250 lb bombs. Beaufighters were not too effective in bombing role, accuracy was of low standard. Similarly, results of rocket projectiles were disappointing. Considerable skill was required to achieve hits with 60 lb rockets.

Obviously, skill was obtained with training and experience, as the RP became an effective weapon with 143 Squadron on the Beau, as well as with other types of aircraft with other squadrons. In regard to a long flight over the sea he does say: 'I made two flights of 8½ hours each over open sea. No auto-pilot! Very tiring!.'

143 Squadron in midsummer 1943 was based at St Eval, and was employed on intruder and long-range fighter interception. On 2 September six of its Beaus took off on fighter interception over the Bay of Biscay, but no prey was sighted. In the afternoon a further patrol of four Beaufighters went hunting over the same area, but this time the hunters became the hunted, for a

---

* Correspondence with author. Retired as Wing Commander J.G. Lingard DFC after a wide range of service.

formation of FW.190s appeared and gave chase. The Beaufighters were not caught and none were damaged.

In October 1943 Leuchars received 455 Squadron and its Hampdens, but in December the squadron was withdrawn from operations so as to re-equip with Beaufighters. Having re-equipped and carried out a training programme, the squadron then moved down to Langham in Norfolk on 2 April 1944, where it joined 489 (RNZAF) Squadron, who were flying TorBeaus. The two squadrons formed the Langham Wing, with 455 (RAAF) Squadron providing the anti-flak escort for 489 Squadron; although its secondary role was armed recce, in which case it went on sweeps armed with two 250 lb bombs and two 500 lb bombs.

By June 1944 there were eight units of Coastal Command operating Beaufighters; these were 143, 144, 235, 236, 254, 404, 455 and 489 Squadrons. With the coming invasion of France by the Allies, the squadrons were relocated at airfields in the southern half of Britain; 144 Squadron, for instance, was based at Davidson Moor. The squadron with 235 and 404 Squadrons was responsible for putting down any enemy shipping approaching the Channel from the west, but from the 1–5 June made no operational sorties.

With the commencement of the invasion on 6 June came the report of three German Seetier

TF.X NE788 of 455 (RAAF) Squadron at dispersal loaded with two 500 lb bombs under the fuselage.

class destroyers sailing into the Channel from the west, so 144 Squadron led by Wing Commander D. Lumsden and fourteen Beaus of 404 Squadron took off from Davidson Moor heading for the French coast, picking up a fighter escort at Land's End. First sighted were six 'M' class mine-sweepers, but these were not attacked as they were not the main target. Then came the three destroyers steaming line astern, so the Beaufighters made their attack out of the sun with rockets and cannon, and appeared to have achieved complete surprise, although LZ180 belly-landed at base on return and NE823 was hit by cannon fire and the observer injured.

Over the next two days the destroyers were attacked by the Beaus, with one destroyer left burning and stationary and an 'E' boat attacked on the 7th, and the other two wrecked by the 8th. On the 9th the last one had run aground on the French coast, so another strike went out, and in conjunction with the Royal Navy made sure that it did not leave. The attacks over the three days going in at 100–1000 ft, and with 404 Squadron's attack on the first day being led by S. Shulemson, now a Flight Lieutenant.

For D-Day operations 143 Squadron had taken up residence at Manston, with 235 Squadron at Portreath and 404 Squadron at Davidson Moor.

From 2 June until the 4th inclusive, 143 Squadron flew a number of anti-E-boat patrols; then on 5 June there was a standby. So that on 6 June 143 Squadron provided cover for the D-Day armada, flying all day on anti-E-boat patrols, twenty-two sorties being flown covering thirty-five flying hours, taking off from 0430–0755 hours with last touchdown at 0925 hours. Then taking off again from 1310–2305 hours and last touch-down at 0100 hours on the 7th. Three E-boats were attacked and sunk and LK972 received bad damage from light flak.

Further attacks were made on E-boats over the next few days, NE764 in the process being hit by a 40 mm shell in the starboard engine nacelle, but made base successfully. This support of the Allied invasion would continue, with attacks on E-boats and other motor vessels over the following month, so that in the end trade got short. Then on 1–2 September, the squadron still operating from Manston, were ordered to attack a main target of about thirty vessels evacuating Boulogne; these split into smaller groups as they proceeded along the French coast, and were harried by naval forces, shore batteries and the Beaufighters.

From these actions boats were left burning and sinking, harried into smaller groups and picked off. NE769 flown by Flight Lieutenant Butterfield and Flying Officer E. Tebb failed to return from combat; whilst NE666 flown by Flying Officer Carr and Pilot Officer Tilley, at the point of dropping their bombs, received a direct hit by a 40 mm shell in their port wing. In damaging the controls this made it very difficult to fly the aircraft; nevertheless, a successful belly landing was made at Manston without injuries.

By now many of the Coastal Command squadrons were back at their home bases operating their own pitches again, and on 10 September 143 Squadron also returned back to North Coates and normal patrols. Then on the 24th fourteen 143 Squadron Beaufighters flew from North Coates to Langham, where the crews were briefed along with 455 Squadron and 489 Squadron for a shipping strike on the Den Helder anchorage.

For this attack Mustang fighter bombers were to act as escort and to attack the flak positions, whilst 143 Squadron were detailed to attack the shipping, and 455 Squadron to make the strikes; 489 Squadron now being the only torpedo

Beaufighter unit in the Dallachy Wing. Twenty-three ships were in the anchorage, and the strike Wing received a great deal of accurate, light flak from *Niueland* and *Terschelling* on the way in. One Beau was seen to dive into the ships and disintegrate, with another one crashing into the sea on fire about one mile outside Den Helder. Squadron Leader D. Pritchard in NT949 with an engine on fire pressed home an attack on one of the merchant vessels, then flew home with holes in the starboard engine and propeller, port inner fuel tank holed, rear fuselage damaged, with half starboard elevator and part of rudder shot away. NE776 and NE973 of 143 Squadron failed to return, as did NT987 of 455 Squadron.

On 4 November 1944 twenty-two Beaufighters of 144, 404 and 455 Squadrons were ordered out on a 'Rover' of the Norwegian coast, with two Mosquitoes of 333 Squadron carrying out a recce ahead of the formation. Landfall was made and the formation proceeded along the coast, with the Mosquitoes investigating the leads and anchorages for any shipping. In Midgulen Fiord two merchant vessels of 3000 tons and three coasters of about 1000 tons were sighted. All aircraft attacked with cannon and RPs, the German freighter *Helga Ferdinand* was hit and sank later, as did a freighter of the Deutsche Levante Line, and a 1000-ton cargo vessel was beached. A box barrage was put up as the formation attacked, but this became spasmodic as the attack developed, with the result that all the aircraft returned safely.

Four days later a further attack was carried out on the Midgulen Fiord, with a formation of twenty-two Beaufighters of 144, 404 and 455 Squadrons led by Wing Commander Gadd and proceeded by two Mosquitoes of 333 Squadron. The formation was flying at 1500 ft, and upon reaching the Norwegian coast, the Mosquitoes went ahead, sighting in the northern end of the Midgulen Fiord two motor vessels of 2500 tons, two coasters of 800 tons each, and one of 1000 tons. Flak started to come up in a box barrage, but this was only light flak initially. As the Beau formation began to go into the attack phase, the second Mosquito called up a sighting of a second and larger force of vessels.

Wing Commander Gadd signalled to the Beaufighter formation to attack the larger force of vessels. Flak then started to come up from ships and shore, with light flak hose-piping up amongst the attacking Beaus, but the attack went in, and as the Beaus pulled away one motor vessel was on fire with black smoke rising to 500 feet, and another was smoking heavily from RP hits. Although some of the aircraft received flak damage they all got back safely.

On 7 December, what appears in hindsight to have been an error in navigation, resulted in a German interception and losses. Twelve Beaus of 144 Squadron, and twelve of 404 Squadron, eleven of 455 Squadron and five of 489 Squadron were detailed for a shipping strike in Allesund Her. Banff Wing Mosquitoes along with twelve long-range Mustangs provided the escort with ASR provided by two Warwicks, Banff Wing in the lead. During the approach to the target the formation was led over Gossen airfield, and from there twenty-five single-engined fighters of the Me.109 and FW.190 type took off to intercept. The strike aircraft attempted to attack but the EA bore in; the Mustangs tried to place themselves between the enemy and the Beaufighters, and the Mosquitoes joined in the fray. The final result was the loss of one Mustang, two Mosquitoes and one 489 Squadron Beaufighter.

By this date, not only were the Allied forces established in France, but moving towards the German border, and Coastal Command squadrons re-allocated to their own airfields again, to resume their normal fields of operation. Some of the Beaufighter squadrons were being considered for, or being re-equipped with Mosquitoes. This would result in the end with 455 RAAF and 489 RNZAF Squadrons moving base to Dallachy, and being joined there in late October 1944 by 404 RCAF Squadron and 144 RAF Squadron, to form the Dallachy Wing; Banff having become an all-Mosquito Wing.

January 1945 found the Dallachy Wing short of 'trade', whilst 9 February was to prove a day of disaster; a combined force of forty-six aircraft was led by Wing Commander C. Milson DSO DFC against a German naval force in Ford Fiord. Upon making landfall an attempt was made to strike at the vessels from the land-side, but with the vessels moored close-in under the fiord cliffs it became necessary to head in from the sea, along the fiord and its shore batteries.

Like so many of Coastal Command's strike operations, the attack went in at low-level to the muzzles of light and heavy flak, accurate light

Rare underwater photograph of 'R' of 248 Squadron, shot down on 12 July 1942, and the wreck discovered off Forsnes, Norway in 1979. (Bjorn Olsen)

flak hosing-up and across like a screen of shooting stars, but far more deadly. Like so many of Coastal Command's strike attacks, it was pressed home with grim determination and raw courage; then the ship anti-aircraft guns opened up and all hell broke loose. Beaus were being hacked out of the sky as they bore in, those that rose from the inferno were pounced on by a formation of twelve FW.190s from Herdla.

The Mustang escort from 65 Squadron attempted to break up the enemy attack, whilst the remaining Beaufighters retreated out to sea, many of them severely damaged with the prospects of a crash-landing at base. Nine Beaufighters failed to return, six from 404 Squadron (RD136, NE761, NT890, NT922, NV292 and NV422), two from 455 Squadron (NV199 and NV196) and one from 144 Squadron ('Y') — it was considered that most fell to the flak, which was a tribute to 65 Squadron's efforts, who lost one Mustang, but claimed three FW.190s shot down and two damaged.

Further attacks were carried out along the Norwegian coastline during the last few months of war in 1945, ships were sunk and Beaufighters lost; sailors and airmen died as the war was played out until the last day. One of these attacks occurred on 8 March 1945, when 455 Squadron with an escort of Mustangs was out on an anti-shipping patrol off Sanday-Svinoy. In Midgulen Fiord was sighted one 4–5000-ton merchant vessel, one 2000-ton merchant vessel, two smaller vessels and a large mine-sweeper. The ships were attacked with cannon and RPs, the flak was heavy and intense, and of all calibres from ship and shore darkening the sky, and enemy fighters came up.

The Mustang escort tried to intervene between the EA and the Beaus, but RD132 was shot down before the Mustang could get the EA. The large MV was by now on fire with smoke pouring from aft of the bridge, the smaller MV on fire and sinking, and one of the smaller vessels sinking — so the job was done for the loss of one aircraft.

The 26 April found the Dallachy Wing led by Wing Commander Gadd with an escort of 19 Squadron Mustangs heading out to Fede Fiord on another ship strike. Again on 3 May the Dallachy Wing led by Wing Commander Gadd were striking; this time the target was an attack in the Kiel Bay area, the scene of the RAF's first big loss of bombers — now the RAF were back. Great quantities of enemy shipping were heading to the few ports still in enemy hands, and there in the Kiel Bay the Wing found the *Java* and *Jutlandia* of about 8000 tons and the *Falstria* of 6990 tons. All were attacked with cannon and RPs, to be left smoking and badly damaged. Four of the Beaus were flak damaged with 'N', crewed by Warrant Officers Brett and Boorer, failing to return.

Towards the end of March 1945, 404 RCAF Squadron moved to Banff and re-equipped with Mosquitoes. 489 Squadron RNZAF flew its last operational sortie on 4 May 1945, with 455 Squadron RAAF on 21 May on the prowl for surrendering U-boats and to be disbanded on 25 May. The Mighty Beau had operated to the end, may she and her Coastal Command crews be long remembered.

Middle East marauding

252 Squadron was formed 21 November 1940, carrying out its first operational sortie on 5 April 1941 (*see* Coastal Search and Strike). This was the unit that was detailed to dispatch fifteen Beaufighters to the Middle East, on detachment to Malta. The decision had been made to either form in the Middle East, or transfer to the Middle East, a number of Beaufighter squadrons. The original transfer of aircraft to the Middle East was by air, the aircraft taking off from St Eval on 1 May; one aircraft returned to St Eval unserviceable, whilst T3229 force-landed in Portugal.

Operating from Malta the surviving thirteen Beaufighters were soon operational, for Malta's survival was also tied up with the North African desert campaign, so operations were carried out over Crete and Sicily, as well as over the coast and seas off North Africa. This resulted in the loss of two Beaufighters which failed to return from operations, whilst a further four were destroyed during enemy raids on Malta. Nevertheless, as in the UK, the Beaufighter had already shown its potential in the strike-fighter role, and demands would increase for the re-equipment of squadrons with the Beaufighter, or the establishment of new squadrons with this aircraft.

The remaining Beaus and their crews had by June 1941 moved from Malta to Edku (Egypt)

R2198 of 252 Squadron over the Middle East camouflaged brown and green down to lower fuselage. (Paddy Porter)

and again 252 Squadron disappears, being incorporated into 272 Squadron, whose CO was Squadron Leader A. Fletcher DFC. 272 Squadron had arrived in Egypt from the UK in May 1941, and had already taken part in a number of operations.

On 21 September 1941 the decision was made to form a Beaufighter night-fighter squadron in the Middle East as soon as possible, the aircraft to be fitted with AI radar, IFF and VHF radio. This was confirmed by Signals War Order 61, which laid down the establishment of the squadron; the aircrew were to be ex-Fighter Command personnel, the squadron to be formed on a mobile basis of sixteen IE Beaufighter Mk. 1s with a reserve of four. It also stated that the aircraft were to be flown out and escorted by Coastal Command Beaufighters. The unit was to be numbered 89 Squadron, the aircraft to be flown out via Gibraltar, and the aircraft only to be flown in the Delta area, so as to prevent the AI from falling into enemy hands — all mention of AI equipment in signals and letters spoke of 'special signals equipment'.

Two days later it was confirmed that two squadrons of Beaufighters were approved for the Middle East, and that from December six aircraft a month would be flown out to cover wastage. 44 Group were given the job of supervising, preparing and training the aircrew for the Middle East Beaufighters, the training to include navigation and long daylight cross-country flights. It had been determined by now that the Beaufighter's safe range was 1350 miles, and 44 Group gave as their opinion, that Beaufighters on the Northern Spain and Portugal route should not be dispatched against a head wind of 5 mph or over. This resulted in the DCAS deciding on the fitment of long-range fuel tanks on all night-fighting and Coastal Command Beaufighters being an urgent requirement.

In October it was announced that the Middle East Command were forming a number of squadrons from their own resources, which included reforming 252 Squadron. With only nineteen long-range Beaufighters available they were converting 252 Squadron to Beaufighters as aircraft and aircrew arrived, forming 89 Squadron later as sufficient Beaufighters and crews became available. The following month the decision was also made to re-equip 46 Squadron with Beaufighters, as the unit had arrived in Egypt without aircraft.

89 Squadron arrived in Egypt in November 1941 and was under the command of Wing Commander G. Stainforth AFC of Schneider Trophy fame. Using its AI-equipped Beaufighters, the first in the Middle East, the

squadron made its debut on 3 January 1942. It was then positioned at Port Said and at night 'readiness' at Gamil.

On 8 April Flying Officer R. Fumerton and Sergeant Bing in X7743 took off from Edku at 0300 hours and were vectored by GCI to an EA at 17,000 ft, the skies clear with bright moonlight. When the contact was made the aircraft was identified as a He.111. As the aircraft closed on the EA it descended to 13,000 ft; so closing in from below, Fumerton brought the Beaufighter to a range of 150 yards and opened fire, pieces immediately flew off the EA. Fumerton's report states: 'We followed him, keeping below and closing in again to 150 yards, gave him another two second burst. There was no return fire, first his whole starboard side was ablaze, then the whole aircraft burst into flames'. After this Fumerton flew the Beau back on to the patrol line, and was vectored onto another EA. This was again identified as a He.111; he closed the range to 100 yards and opened fire, almost immediately there was an explosion, and the EA in flames went down in a spiral dive.

252 Squadron reappeared again as a complete squadron in January 1942, and along with 272 Squadron were in the thick of battle during the next few months. 272 Squadron provided escorts for different types of aircraft and different squadrons, as well as strafing MT columns and airfields.

Some of the better-known exploits of 272 Squadron took place in November 1941. On the 20th of the month eight Beaufighters of the squadron attacked the *Luftwaffe* on the ground and in the air, with three of the pilots shooting down four Ju.87s and a Fiesler Storch, and then destroying fourteen Ju.87s and an Me.109 on the ground. On the following day, which was the fourth day in succession, the squadron strafed first Tmimi and then Matuba, destroying on the ground three Ju.87s, one Ju.88 and one Hs.126 — all without loss.

On 6 December 1941 the Squadron sent four Beaus to attack Tmimi again; but Pilot Officer Snow and Sergeant W. Dutton in 'J' failed to return (Obt Rodel of II/JG27 claimed a Beaufighter), 'J' crash-landed with port engine on fire about two miles north of Tmimi airfield. A further four Beaus then took off at 1115 hours and attacked Tmimi airfield, destroying five Ju.87s on the ground. Flight Lieutenant

Campbell in 'O' after being hit by flak force-landed about forty miles south of the target area; Pilot Officer Hammond in 'P' landed beside him and after a repair both aircraft took off again. Campbell was forced to land again, but this time nothing could be done, so again Hammond landed and picked up the crew, then shot up 'O' and returned to base. 'H' flown by Pilot Officer Stephenson and Sergeant Olive, was hit by ground fire, with the pilot hit in both legs and the observer badly injured. A belly landing was carried out at Edku, but the observer died of a serious head wound.

The priority during May 1942 was interception of Axis aircraft and supplies. On the 12th five Beaufighters of 252 Squadron with nine Kittyhawks of 250 Squadron were detailed on an anti-Ju.52 patrol. T4868, T4896, T4831 and T5028, led by Squadron Leader Wincott and Sergeant Kilminster in T4881, took off from Gambut satellite airfield at 0805 hours and intercepted thirteen Ju.52s escorted by a Me.110 just after take-off from Maleme, bound for Derna. Nine Ju.52s and the Me.110 were shot down into the sea or onto the shore. T5028 flown by Sergeants Cripps and Bateman, was hit in the wing root and caught fire; the aircraft touched the water with the left wing and went in.

On 26 June 1942 Squadron Leader P. Evans and Sergeant Houston of 89 Squadron flying X7754, were on convoy patrol, when they were warned of an EA shadowing the convoy. The observer saw the EA flying low down and moving towards the convoy, so warned the pilot, who dived the Beaufighter towards the EA, which was identified as an Italian SM.79. The Beau came up behind the SM.79, which had dropped a torpedo towards the convoy, and gave it a few bursts of gunfire, the second burst affecting the rudder. X7754 now travelling at 240 mph and about twenty feet above the sea, poured more gunfire into the SM.79, which returned the fire, then: 'I then closed in to approximately 50 yards and gave a long burst of machine-gun fire, my cannons having stopped, which set his starboard engine on fire, and a lot of pieces fell off'. The SM.79 crashed into the sea about ten miles north-east of Port Said.

Another anti-Ju.52 patrol was mounted on 11 July, when ten Beaufighters of 272 Squadron and three of 252 Squadron took off on an interception. Between Crete and Tobruk a formation of thirty Ju.52s, one Ju..88, one He111 and one Me.110 were met. Squadron Leader Ogden leading one section shot down one Ju.52 and damaged another, before he was attacked out of the sun by the Me.110, strikes being made on the Beaufighter. Surprisingly enough only three Ju.52s were claimed as destroyed and three others damaged; the troops in the transports firing from the windows in conjunction with the escort aircraft resulted in two of the Beaus being damaged, one of which crash-landed at base.

One of the pilots said of the Beau:

'It was so well constructed that it absorbed punishment like a sponge, initially the cockpit was a little awe-inspiring, but that turned out alright. In the Middle East we did have problems flying on one engine, there never seemed to be enough urge.'

Like all engines the Hercules lost power in the hot climate of the Middle East, with, as on the Spitfire and Hurricane, further power loss and increase of temperature due to the air filters fitted to clean the air to the carburettor. 'If a satisfactory type of bypass of the filter had been fitted to the air intake, some of this loss of power could have been regained once in clear air in flight.'

In regard to the Middle East situation, the *Luftwaffe*, similar to the British, about this time set up a Mediterranean Air Command under the command of Marshal Kesselring, and further subdivided to the following *Luftwaffe* Generals: Lörzer in charge of the units that operated from southern Sicily; Geisler commanding the *Luftwaffe* formations — in particular the dive-bombers — stationed in Sicily, and Hoffman von Waldau as *Fliegerführer Afrika*. The latter appointment indicating the importance that the Germans were at last giving to this area.

The British in North Africa were suffering from Rommel's activities and Malta was reeling from two months of concentrated air attacks; so as well as the need for Spitfires for defence, there was even more need for Beaufighters to take the fight to the enemy. With the *Luftwaffe* and Italian Air Force neutralizing Malta as a strike base, more Axis shipping was getting through to North Africa, and more formations of transport aircraft were flying troops and supplies to Rommel, who was now attacking the British forces at Gazala and would force them back to El Alamein.

So the squadron's activities over the coming weeks would be concerned in attacking the landing grounds in North Africa, striking at the air transport formations and making shipping strikes or providing escorts for Beauforts and Hudsons on shipping strikes. 252 Squadron on 2 July for instance, sent out sorties to strike at LG16, LG17 and 18 at Fuka, where three

T5037 of 272 Squadron missing in the desert, 19 December, 1942. (Mathieson)

Me.109s were destroyed on the ground and one damaged. Then one Beaufighter was jumped by a Me.109 and suffered repeated attacks, the Beau pilot carrying out evasive action as close to the deck as possible. The Me.109 then broke off the attack and formated alongside the Beau, the enemy pilot waved to him and left — presumably out of ammunition; whilst the Beaufighter, badly shot up, managed to return to base. From the day's sorties however, two Beaus flown by Flight Lieutenant Malcolm with Pilot Officer Sharp and Pilot Officer Ydlibi with Sergeant Spiller failed to return.

At this time 89 Squadron were based at Abu Sueir and were providing night cover for the Nile Delta, sometimes operating from Abu Sueir, Port Said and Edku. The night of 4 July was to prove productive, for large numbers of the squadron's Beaufighters scrambled that night from both base and Port Said. This resulted in Wing Commander Stainforth and Pilot Officer Lawson in X7704 destroying a Ju.88; Flight Lieutenant Waddingham and Pilot Officer Cumbers in X7719 destroying a Ju.88; Flying Officer Harding and Flight Sergeant Horspool damaged a Do.217, and Sergeants Pring and Phillips in R2266 damaged a He.111.

At Edku 272 Squadron received its first Beaufighter with belt-fed cannon on 25 July 1942; this was usually flown by Pilot Officer Palmer. Unfortunately, snags were still being experienced with the cannon installation, as related by George Lovett, who served with 272 Squadron at that time:

I experienced a lot of snags with the 20 mm cannon, mainly because of the design of the cartridge chutes, which tended to block with empty cartridge cases, causing the guns to jam. Eventually, by increasing the size and angle of the chutes, the problem was overcome. But in the early days the pilot used to curse on his return, having lined up a target, after a short burst from the guns they would jam.

Apart from its cannon installation problem, the Beaufighter was a most respected aircraft, for it had a large and sturdy undercarriage, so was able to use the rough airstrips, so common to the Middle East. Many of these became quagmires during storms, but the Beau could plough through most of them, where other aircraft became immobile or stood on their noses. Its

opponents at night were mainly the He.111 or Ju.88, although the SM.79 and Cant Z.1007s were also encountered. Against the He.111 the Beaufighter had a large overtaking speed, but against the Ju.88 the differential was small, so this and greater manoeuvrability could make the difference between life and death of the Junker's crew.

Warrant Officer Thomas and Sergeant O'Toole in X7628 of 46 Squadron on 27 July, scrambled from LG224 on a night interception. O'Toole obtained a contact at 15,000 ft, which was followed by a visual of the target at 1500-yard range. Thomas closed the target to 300 yards on the EA's port side and identified it as a He.111; closing to 150 yards and opening fire, strikes were seen on the port wing and engine. Return fire came from the top dorsal gunner, but closing in to 75 yards dead astern the second long burst hit the fuselage and wing root and the engine started smoking. The EA then went in to a steep dive and crashed near Gebel Gatrani.

On 21 August Wing Commander G. Reid and Sergeant Jackson in X7779 of 46 Squadron were scrambled against 'bandits'. A contact was made, then a visual, which identified the EA as a Ju.88, dead ahead at 400 yards. Reid slowly closed the gap to 100 yards dead astern and gave a two-second burst. The EA immediately commenced evasive action with diving and turning, Reid following gave a number of short bursts that hit the port engine and fuselage, then the starboard engine. The Ju.88's port engine was now well on fire, and the aircraft in a dive, continued this into the sea.

On 10 August 252 Squadron sent four of their Beaufighters on detachment to Cyprus and whilst proceeding two of them sighted a Ju.88. Pilot Officer E. Derrick in T4940 and Flying Officer Reed in X7633 moved in to the attack, when the EA made a steep turn that brought it into the line of approach of Reed's aircraft, which gave it a long burst of cannon fire from almost head-on and above. The Ju.88 immediately broke up in mid-air and crashed into the sea.

Malta was not keeping quiet either, the 89 Squadron detachment there had the AOC, AV-M Keith Park, as a passenger on 19 August. It was so typical of this Commanding Officer, who always wished to see for himself, and Flight Lieutenant Edwards in V8268 flew him around Malta to check the searchlights and their accuracy in intercepting aircraft. Following this, on the 26th of the month the AOC gave the squadron permission to fly Intruder patrols over Sicilian airfields and seaplane bases, and to strike at Sicilian shipping. On the same night Flight Lieutenant H. Edwards and Pilot Officer J. Ross flew to Sicily, intercepted and shot down a Dornier 18 flying-boat, and then proceeded to shoot up an airfield flare path and Chance Light.

September 1942 was to see a turnabout in the fortunes of the *Afrika Korps* and the start of the end of the Axis Armies in North Africa. A small, wiry and unknown British General named Montgomery and his debonair chief, General Alexander, had taken over in August; new plans were made, more material poured into the British forces, more training undertaken, and more important still, the Army General Commanding and his RAF opposite number had their headquarters together for the first time. The 30 August saw Rommel's attack commence on Alam el Halfa; by 2 September Rommel's armies were being sucked into a cauldron, his supply lines and troops — and himself — attacked by RAF bombers and fighters. Off Tobruk his fuel tankers were being sunk once more — the retreat would begin.

During the September 89 Squadron detachment made attacks on Syracuse, and on the 6th provided four Beaufighters as top cover escort for a Beaufort and Beaufighter shipping strike. One of the Beaus failed to take off, so leaving only three to do the escort work; however, these justified themselves during the strike by shooting down a Ju.88, and damaging a Macchi 200 and a flying-boat.

In all of this 252 Squadron Beaus were on offensive strikes destroying ships and barges, and strafing road traffic. 272 Squadron were doing the same, as well as providing escorts for shipping convoys. Yet on 3 October, Rommel was in Berlin and still talking about entering Cairo — 252 Squadron did not know this, and sent on the 7th four Beaus to attack the seaplane base at Menelaio Bay. Upon arrival they found four CZ506s at moorings and five other aircraft on the slipway, so it was business as usual, attacking the moored aircraft and one on the slipway before making their way along the coastal road. There they destroyed a lorry towing a fuel bowser, a lorry with oil drums and strafed

252 Squadron Beaufighters attacking Greek island base for seaplanes. (Paddy Porter)

a lorry of troops. Next to get strafed were a parade of troops, who were caught unawares and shot up. Turning for home they next found eight F.-boats out at sea, so went in to the attack, but this time flak was intense and T5114 was damaged, but made base and crash-landed.

The same squadron was in luck again on 25 October, after a series of offensive sweeps, escort to shipping convoys, and road strafing. Seven of the squadron's Beaus were out on a search for an enemy convoy that was reported in the area of Tobruk. When found the convoy comprised one 800-ton petrol carrier and a torpedo boat, with an air escort of one Do.24 flying-boat and three Ju.88s, indicating Rommel's heavy requirements for more and more fuel for his Panzers. So while three of the Beaus dropped down to deal with the ships, the other four took on the enemy aircraft. Within minutes the Beaus were speeding away, with both the enemy vessels wreathed in black smoke, the Do.24 was set on fire and crashed into the sea, and two of the Ju.88s shot down.

In the same way 272 Squadron were also game hunting, flying anti-Ju.52 patrols. Then on 12 November Squadron Leader A. Watson led six Beaufighters along the coast and found south-east of Pantellaria six Italian SM82s. Attacking the unescorted transports the Beaufighters soon put all six into the sea. In the same month Squadron Leader R.G. Yaxley was promoted Wing Commander and took over as CO of the squadron.

Also in November 600 (City of London) Squadron was ordered to move to the Middle East; this was an established night-fighter squadron with an enviable record, and seventeen of its Beaufighters would fly direct to North Africa via Gibraltar. 255 Squadron was another unit ordered to fly direct to North Africa for night-fighter duties there, transiting and refuelling at Gibraltar, before flying on to Maison Blanche. The AI radar equipment of both squadrons' aircraft was removed before departure.

With operation 'Torch' planned for 8 November 1942 with Allied landings to be made in the Algiers area, the plans were for the RAF to provide most of the air defence of the supply ports, air support for the troops and tactical co-operation with the Army; the strategic bombing operations being in the hands of the USAAF.

Amongst the fighter defences of the area were two Bristol Beaufighter squadrons as night fighters; these were 255 Squadron commanded by Wing Commander D.P. Kelly and 600 Squadron commanded by Wing Commander J.R. Watson, both squadrons flying Mk. VIF aircraft. Two days prior to the landings in North Africa sixteen Beaus of 272 Squadron were moved from Edku to Takali in Malta — these were to provide further support to the Allied forces.

The initial landings around Algiers proceeded quite well, in some cases the strongpoints were captured without any problem, but the French Air Force (Vichy) put up resistance and losses were suffered on both sides. The landings on the beaches at Oran proceeded satisfactorily and only ground to a halt when the American troops tried capturing the harbour there. Likewise at Casablanca, there was initially success, with the Americans under General Patton surging inland with the assistance of the naval guns, but the inexperience of the Americans soon brought delay, until aircraft from the USS *Ranger* beat down the opposition. In all cases, both Allied and French air forces suffered losses.

On 15 November 255 Squadron Beaufighters arrived at Maison Blanche, the first night fighters to arrive, but without their AI radar (removed in the UK for security reasons) their role was diminished. Prior to this on the 12th, Beaufighters of 272 Squadron had patrolled the Tunis-Sicily area; the section of seven aircraft was led by Squadron Leader A. Watson and intercepted a formation of six Savoia SM75 transport aircraft, five in German markings, and one in Italian. The six aircraft were claimed as shot down, but the Italians only recorded the loss of five.

Maison Blanche airfield was crowded with American and French aircraft, and little space was allowed for 255 Squadron aircraft. This prime target was soon taken advantage of by the German bombers, and on 16 and 18 November raids destroyed a number of Allied aircraft, leaving 255 Squadron with about three serviceable aircraft. 600 Squadron Beaus arrived at Blida on the 18th, but it was not until 29 November that the unit opened its North African 'score-card'. Then on 7 December the squadron moved to Maison Blanche, with a new CO taking over later on in the month; this was Wing Commander C.P. 'Paddy' Green.

GCI control was fully mobile and transportable to any point, the unit being under the control of Squadron Leader Brown. The fully (!) tropicalized Beaufighter VIs with their power-losing air filters could just about make 22,000 ft — with luck and a push — so it was most fortunate that at this period of establishing the 1st Army in North Africa, that the night bombers in general did not fly above this height when attacking Allied bases; meanwhile the AI radar equipment had at last arrived by sea and all haste was made in refitting it.

Both 255 and 600 Squadrons were called upon at times to fly standing patrols off Sardinia, as well as the night-fighter scrambles. While 272 Squadron on 24 November found an interesting target, during a day of carrying out a series of sweeps along the Tunisian coast and the Gulf of Tripoli. On one sweep, taking off at 1300 hours, three Beaus were detailed to recce the eastern Tunisian coastline, when at about fifty minutes after take-off Flying Officer Coate and Flight Sergeant E. Miles in T5038 sighted a Bv.222 six-engined flying-boat thirty miles off Linosa.

The attack is described in the squadron ORB as:

> After a chase of 20 miles Flying Officer Coate made a beam attack out of the sun. The first short burst knocked large pieces from the fuselage and upon closing in he attacked the port engines and petrol tanks, which were set on fire with a long burst. The aircraft began to lose height and hit the sea. It bounced about 60 feet into the air, the left wing dropped and the aircraft went into the sea in a half roll and blew up.

The aircraft was the Bv.222 V-6 of *Lufttransportstaffel* 222, and was the first of its type lost in action.

Coate continued his patrol and almost immediately sighted a Do.24 flying-boat forty-five miles off Linosa. He carried out an attack and strikes were seen on the fuselage which knocked pieces off rudder and port wingtip. The EA fired off smoke puffs, presumably to attract the attention of some Me.109s at about 2–3000 ft which were circling a hospital ship. With this the three Me.109s turned towards the contest, and so the Beaufighter withdrew from the scene.

In spite of bad weather affecting both the land battle and reducing the effectiveness of air support, RAF units kept on flying. 272 Squadron

almost lost a Beau flown by Pilot Officer Stead to a Me.109, when carrying out a recce of the Gulf of Tripoli. The Beau was hit in the tail, starboard engine and oil tank, and the port wing; in spite of this Stead managed to throw off his attacker and returned to Malta to a crash-landing.

Malta-based 227 Squadron, a day-fighter unit, was another squadron that combined its offensive sweeps with ground-strafing of enemy airfields and ground targets. Like 272 Squadron they also totted up a fair share of enemy aircraft destroyed in air combat. The squadron had been taken over by Squadron Leader A. Watson in December — this officer had, while flying with 272 Squadron, been shot down during ground strafing, and with his navigator had walked back to the British lines.

At the end of 1942, 46 Squadron moved from Edku in Egypt to Malta to join the other Beaufighter squadrons in the area, and soon found plenty of 'trade' as the North African campaign drew to its end in the first few months of 1943, and Kesselring threw in more aircraft from Italy and Sicily; even armed giant Me.323 aircraft were used and shot down. Offensive sweeps were flown by the Malta-based squadrons, sometimes escorted by Spitfires of 229 Squadron.

The AOC chose 272 Squadron to supply Beaufighters for his transit visit of the Middle East. The take-off was made on 27 December, with the AOC flying in EL274 crewed by Pilot Officer Cleggett and Sergeant Watkins; an escort aircraft was flown by Wing Commander J. Buchanan and Sergeant J. Orchard in T5112. During the flight Buchanan sighted two FW.200s in the distance, so moved in to the attack; on being sighted the two EA closed in to tight wingtip formation. Making four attacks from the rear, strikes were seen on the one aircraft's outer port engine, but concentrated fire from the two aircrafts' three gun positions damaged Buchanan's aircraft, and he was forced to pull away.

With that the AOC ordered Cleggett to attack as well, but as they dived in the two FW.200s commenced to climb up and subjected the Beau to intense return fire; and although it started to climb it could only get to 3500 ft. Then the port engine vibrated and seized, so after feathering the propeller both Beaus returned to Malta, which they made successfully — and the two

Bob Atkinson, engine fitter with 252 Squadron, takes a breather with his Beaufighter 'out in the blue' at Gambut. (Paddy Porter)

Fw.200s went on their way! Later on in the day the AOC flew off again, Buchanan again as escort, this time without incident.

During December 252 Squadron had a detachment at Berka 3; however, in January 1943 the squadron came together at Berka, and was mainly engaged in flying escorts to shipping convoys. By February the squadron had moved again, this time to Magrun, a new CO was also appointed, Wing Commander P. Ogilvie; the previous CO, Wing Commander Bragg, had been shot down during a ground strafing sortie and become a PoW.

The start of 1943 saw the Axis forces pushed out of Libya for the last time and into Tunisia, the British 8th Army moving into Tripoli and westwards across North Africa. The main fighter unit in the area was *Jagdgeschwader* 77, with three *Gruppen* and commanded by Major J. Muncheberg; the three *Gruppen* flew Me.109F and Gs. On the British side the air superiority role was in the hands of 244 Wing, comprised of four Spitfire squadrons and one Hurricane

squadron. Night fighting was in the hands of 89 Squadron, night intruding the province of 46 Squadron and 252 Squadron carried out day-fighter strikes; all three squadrons were equipped with Beaufighters.

On 13 January four Beaufighters of 153 Squadron were detached to operate from Tebessa; however, on arrival they were attacked by USAAF Warhawks, which resulted in one being shot down, so the remaining three were based at Youks-les-Bains. The detachment was fated, for two of the remaining three were then damaged by a ground-strafing attack by FW.190s; so on the evening of 15 January the sole serviceable aircraft rejoined 153 Squadron back at Maison Blanche.

From Malta 272 Squadron carried out further day sweeps and strikes; on 16 January 1943 four Beaus left on an offensive sweep of the Lampedusa and Tunisian coast. Flight Lieutenant Coate and Flight Sergeant E. Miles in T5171 sighted west of Malta a Ju.88, and giving chase Coate caught it south of Lampedusa. Approaching from underneath Coate opened fire from 150 yards at the starboard engine, which caught fire; he then fired at the ventral gun position, followed by the port engine. The EA then began losing height, crashed into the sea and exploded.

Two days later a 255 Squadron Beaufighter flown by Flying Officer Gloster and Pilot Officer Oswald was patrolling over Bone, when they were vectored to 20,000 feet on to a Cant Z.1007. Gloster opened fire from 600 ft, seeing strikes on the fuselage, then the EA exploded. A few hours later, another of the squadron's Beaus crewed by Flying Officers Greaves and Robbins closed to 300 ft astern of a Z.1007, but after a short burst three of the cannons jammed. The EA went into a steep dive with Greaves following and Robbins trying to clear the cannon. Then at 8000 ft with the cannon cleared, Graves poured in a burst, the three engines and fuselage were hit and caught fire, and the EA went into a vertical dive.

As was to be expected in the desert of North Africa, serviceability of the aircraft was a problem and spares at a premium, with the continual movement of the ground battle hardly helpful to supply. A further problem also occurred as related by George Lovett:

> 28 January 1943, 272 and 252 had an embarrassing experience during a night-time tropical storm. The following morning most of the starter trollies were found chocked up with their wheels missing. Apparently, during the storm the Arabs had come through the perimeter fence with donkeys and calmly dismantled the lot!

Lovett considered the Beaufighter ideal for the role that it was called on to perform in the Middle East, and remembers 272 Squadron's method of strafing ground targets as most effective '. . . was to fly up the desert by flying over the sea. Then to swing in over the desert road and travel in the opposite direction from which the enemy was expecting an attack, with terrific success.'

The Beaufighters along with other Allied aircraft were by now searching and striking, by day and night, ship and aircraft convoys bringing succour to the Axis forces in North Africa, which were being squeezed between the 1st and 8th Armies. The night of 14 February 1943 was to show a little trade for the night fighters; 1900 hours found a 46 Squadron Beau flown by Flight Sergeant Holmes and Flight Sergeant Bell scrambling from Tobruk. Climbing the aircraft to

Some of the ground crew lads of 272 Squadron.
(George Lovett)

18,000 ft over the Abu Amud area a radar contact was gained; this turned into a visual at 19,000 ft, when Holmes identified a Z.1007bis. Closing in to 350 ft he poured in five bursts of cannon, at which the bomber burst into flames, staggered and then dived vertically into the sea.

Flying Officer Sage and Flight Sergeant Cockburn took off five minutes after Holmes and at 18,000 ft got a contact below them; slowly losing height a visual was obtained at 15,000 ft when another Z.1007bis was identified. Sage fired at 200 ft range and saw his fire go into the wing and fuselage of the EA, the starboard engine immediately burst into flames, the aircraft then appeared to burst into flames and dived vertically through the clouds. The same night II/KG76 lost a Ju.88, when Flight Sergeant Blackwell and Sergeant Melrose of 89 Squadron obtained a contact and visual on an EA. Blackwell poured a burst of cannon into the EA over Tripoli, then after a stern chase a further burst of fire, this time the aircraft dived away and crashed.

On 17 March 1943 nine Beaus of 272 Squadron led by Wing Commander Buchanan took off around 1130 hours to provide an escort for Beauforts of 39 Squadron, to make a shipping strike on an enemy convoy of two motor vessels and three destroyer escorts off Point Stelo. Five minutes after take-off T5174 had engine trouble and was forced to ditch, the crew being picked up by launch. The convoy was sighted at 1325 hours and was seen to have an escort of ten Me.110s and Ju.88s, a Ju.52, a Do.217 and a He.115. The Beaufighters engaged the air escorts to allow the Beauforts an uninterrupted run-in — apart from the flak!

At this time four Me.110s of III/ZG26 were in transit flight from Trapani to Gerbini and saw the attack take place. So dropping their long-range tanks the four Me.110s dived in to the attack and achieved total surprise, claiming four or more aircraft shot down — but only Beau X8132 flown by Sergeants L. Schultz and Wainwright failed to return.

The Beaufighters however suffered with cannon trouble and this resulted only in damaged enemy aircraft. Buchanan in EL274 hitting a Do.217, Squadron Leader Coleman in EL323 damaging the Ju.52 and He.115, and Flying Officer Welfare in T5038 damaging a Me.110. Apparently, early on, an investigation into stoppages found that the cause was due to armourers greasing the cartridges with a mixture of petrol and oil to ease the movement of the cartridge rounds in the drum, but in some cases this mixture had found its way inside the round, rendering the propellant unusable. Whether this was again the case the squadron record book does not state.

At this period 153 Squadron was based at Maison Blanche, Algiers, and was commanded by Wing Commander Moseby DFC. It had commenced operations from this base in late December flying night sorties over the Algiers-Bone area, but did not open its score until 13 January 1943, when V8631 flown by Flying Officer Williamson and Sergeant Lake shot down a Piaggio P108B (identified and claimed as a FW.200). Another aircraft of the same type was shot down by Flying Officer Haddon of the same squadron, flying his first operational flight in Africa; the attack going in at about 11,000 ft.

The evening of 11 May 1943 found Flying Officer Stephenson and Flying Officer Hall in EL174 scrambling from Djidjelli on interception. A vector was given them by GCI, and about ten Ju.88s passed overhead. In turning to follow them visual contact was lost, however N/RO Hall got a contact on an EA and directed Stephenson onto the target — a Ju.88, which he attacked from 800 ft firing a two-second burst, and saw the EA (of III/KG76) burst into flames and dive into the sea.

As he was now approaching close to the harbour and its adequate AA defences, Stephenson turned away out to sea, and in the process picked up the track of another Ju.88, which was intent on racing home having delivered its bomb-load. However, a visual on the EA was lost and Hall had to obtain a contact. With a visual again obtained the Ju.88 then started to jink violently and its gunners fired back. Speed had now increased to 290 mph as the EA lost altitude; so diving to within 50 ft of the sea Stephenson got below the EA, and at a range of 900 ft gave it a full two-second burst, causing the Ju.88's starboard engine to burst into flames, and the aircraft then dived into the sea.

The Mediterranean Air Command at this period was under Air Chief Marshal Sir Arthur Tedder, who controlled three air forces, two of which were North-West African Tactical Air Force and North-West African Coastal Air

Force, both of which at some time or other employed Beaufighter squadrons. 323 RAF Wing had 153 Squadron and 255 Squadron detachment; 324 RAF Wing had 89 Squadron detachment, 255 Squadron and 600 Squadron; while in Malta were 46, 227 and 272 Squadrons.

This Middle East Command held a number of outstanding Squadron Commanders and pilots; some of the latter were NCOs, and although it is invidious and unfair to many others to name some, not to mention the outstanding ones is just as unfair. Wing Commander J.K. Buchanan* must surely be the most outstanding Beaufighter intruder pilot, at this period in command of 272 Squadron; with Squadron Leader F.D. Hughes of 600 Squadron starting to emerge as the most successful night fighter on the Beau.

By June 1943 preparations were in hand for the invasion of Sicily, named 'Operation Husky', with the result that the Allied-occupied ports in North Africa were holding large quantities of war material. These were would-be targets for the *Luftwaffe*, with the result that night raids by Ju.88s and Z.1007s increased. To help in the night defence, 219 Squadron under the command of Wing Commander A. McNeill Boyd, was posted in to Bone, and became operational on 30

252 Squadron Beaufighter at Edku; pilot looks surprised that it's so clean! As blast tubes still sealed over, it could be a new delivery. (Paddy Porter)

June. Sicily was invaded on 10 July 1943, and although aerial opposition was slight for two nights the scene soon changed; but this made good hunting for the night intruder Beaufighters from Malta, whilst the TorBeaus were hunting for enemy shipping.

In the Middle East by May 1943 there were ten squadrons equipped with Beaufighters, shortly afterwards being joined by a detachment of 256 Squadron at Malta, and 219 Squadron moving to North Africa in June 1943. So that by the end of July the following squadrons were based in the Middle East area, 46, 89, 108, 153, 219, 227, 252, 255, 272, 600, 603 and 256's detachment. This was followed by the re-equipment of 39 and 47 Squadrons with Beaufighters, and the entry into the area of 144 Squadron. These three latter squadrons were soon in operation and achieving remarkable success around the Sicilian, southern Italian and

* During his second tour on Beaufighters in 1944, this officer was shot down, and by the time he was found had died of thirst and exposure.

Sardinian coasts, with considerable shipping being sunk by torpedo and cannon prior to the invasion of Sicily.

Operating from Malta was 600 Squadron with their Mk. VI night fighters, 108 Squadron with their Beaufighter night intruders and 272 Squadron with their Mk. VI strike fighters. The NW African Coastal Air Force was meanwhile operating 219 Squadron and 153 Squadron as night fighters, and 39, 47 and 144 Squadrons as ship strikers with their Beaufighter TF.Xs. Air Defence Eastern Mediterranean had 46 Squadron under 219 Group and a 108 Squadron detachment under 212 Group.

During the Sicily invasion night-fighter squadrons were having a most successful time; on the night of 14/15 July four Beaufighter pilots claimed thirteen bombers downed. Wing Commander C. Green of 600 Squadron accounted for four EA and Flying Officer J. Turnbull of 600 Squadron accounted for another three. So that by the end of July the score of the squadron over the month was twenty-five.

Beaufighters of 227, 252, 272 and 603 Squadrons were, during this period, pushing their sweeps into the coastal areas and up the Ionian Sea as well as the Aegean Sea. On the 17th they carried out an armed recce of the west coast of Greece. This was a softening-up process ready for the planned invasion of the Aegean islands — a plan which would founder with unsufficient forces being committed, and with the decision of the Germans to defend and reinforce the area.

As has been said before, the Beaufighter was not an easily manoeuvred aircraft, and certainly not the type to take on single-engined fighters, yet sometimes evasive action combined with aggression could win the day even in such circumstances. On 25 July two 39 Squadron Beaus were detailed to fly to Pianosa Island and drop packages; on the approach they were intercepted by two Me.109s, so both Beaus turned into the attack, Flying Officer Cullen shooting one down in the process.

The Aegean area was during 1943 to prove an expensive area for aircraft, and in coverage provided for the destroyers operating around there and escort for convoys, 133 aircraft were lost during the campaign, including fifty per cent of Beaufighters committed. One of the units called on to provide offensive patrols and escorts was 600 Squadron. The commanding officer

during this period was Wing Commander Chater, and the squadron often used bombs as well as cannon to quell any opposition, operating at this time from Berka III.

On 15 August 1943 an offensive sweep by two aircraft included Wing Commander Chater, and was carried out over the Peloponnese, with bombs being dropped on Kalamatha landing ground and the harbour flak posts cannoned and silenced. This was followed on the next day by four Beaufighters led by Squadron Leader Atkinson in an attack on shipping in the harbour of Missalonghi in western Greece. A direct hit was made on a 3000-ton motor vessel as well as two near misses, then a cannon attack was made on that and another vessel, and seamen were seen to dive over the side into the water during the attack. Before leaving the harbour, the Beaus did a quick tour and cannoned the flak posts.

Convoy escort work and general flying were carried out, intermingled with offensive patrols along the Greek coast and across the Aegean. Then on 24 September, after being escorts to convoys during the morning, four Beaufighters took off near midday on a sweep for enemy transport aircraft evacuating Corsica. One of the aircraft had to turn back as it became unserviceable, then at 1421 hours the other three Beaufighters sighted six Ju.52s on an easterly course down at sea-level. Wing Commander Chater leading the pack destroyed one and damaged another, while the other two Beaufighters got one each.

However, JL731 flown by Pilot Officer Megone and Pilot Officer Williams, was hit by return fire and the port inner fuel tank was on fire, so Megone headed for Pianosa Island and made a successful landing. The fire was put out, the Italian population proved friendly and helpful, and warned the crew that the Germans visited the island every evening. So the decision was made to chance a take-off and, hopefully, a return to base. Once airborne Megone made for Corsica and crept along the coast, down Sardinia and landed at Berka III at 1830 hours — about twelve hours after take-off! All the aircraft had bullet holes in them, and one had to carry out a wheels-up landing due to damaged hydraulics.

The war had now, for the Anglo-American forces, reached mainland Europe, for on 3 September 1943 the British had landed two Divisions of the Eighth Army at Reggio di

Two 29 Squadron TF.Xs on a shipping strike with the right-hand one fitted with a Yagi-type ASV nose aerial.

Calabria in southern Italy, and by the 8th it was known worldwide that Italian Marshal Badoglio had already signed Italy's surrender to the Allies. The following day the USA Fifth Army made a seaborne invasion at Salerno, so the night-fighter Beaufighters were out over the invasion area looking for German night bombers.

On 14 September 46 Squadron was involved in the take-over of the island of Cos, Squadron Leader W. Cuddie and Flight Sergeant L. Coote in V8578 in company with Flying Officer Atkins and Flight Sergeant Mayo in V8637, flew special equipment over to the island. Cuddie landed while Atkins provided cover. A week later the squadron started changing its night-fighter Beaufighters for the coastal type, and a detachment was sent to Cyprus on night intruder operations over Rhodes.

The Germans were not to accept this take-over of the islands in the Aegean, and on 3 October began an attack on Cos with their sea and air forces in strength. This was contested by the British, but with their forces committed in Italy and few airfields available in the area, the Germans were soon making progress. 46 Squadron was one of the units contesting the area, and Wing Commander J. Read with Flying Officer Peasley in JL913 led seven of the squadron's Beaus against a convoy that included

two destroyers and three motor vessels. Read's aircraft was damaged by flak and crashed into the sea off Turkey; Read was killed, but his Nav.RO was injured and picked up.

Squadron Leader Cuddie was seen going in low on the water into the flak barrage from the ships, then his aircraft was struck, staggered temporarily and exploded. Two other Beaus were ditched after being hit by the intense flak barrage from the ships. Strikes were seen on the destroyers and motor vessels, one Ju.87 was shot down and another damaged. By the following day Cos was again in enemy hands.

During October 89 Squadron was withdrawn from operations and received an order posting it to the Far East, still under the command of Wing Commander W. David DFC AFC. As a night-fighter squadron 89 had achieved some good scores in the Middle East, and in particular during the early days during the Sicilian and Italian invasions by the Allies. Now the Allies had all southern Italy occupied with the Axis troops still on a fighting retreat.

A South African Air Force unit, 16 Squadron, had been flying Beauforts in North Africa, but on 1 January 1944, whilst based at Berka III, had

115

*A Junkers Ju.52/3 floatplane target for a 603
Squadron Beaufighter over the Aegean.*

commenced re-equipment with Beaufighters;
unfortunately the airfield was in general
unserviceable, so little training could be done
from there, which entailed training away.
Practice with both cannon and rocket projectiles
(RPs) was carried out, and an enthusiastic
training programme embarked on. Tragically, the
first casualties with the Beaufighters occurred
during practice flying, when on 4 February two
of the aircraft suffered a mid-air collision;
although both aircraft recovered from the
collision, one aircraft at first seemed to be under
control, but then dived into the ground near to
the Benghazi road. The other managed to recover
from a dive and returned to base.

The squadron's first official operation with its
new aircraft took place the following day, when
it provided long-range escort to a convoy. A
similar task was performed the following day, in
spite of bad weather. The first offensive
operation by the squadron took place on
7 February, when the CO, Lieutenant-Colonel
Lorentz led four Beaus on a sweep following
the southern coastlines of the Peloponnese
and covered the Gulf of Lakonia and
Messina. Although each aircraft carried eight

25 lb armour-piercing RPs, no targets were
found.

On the 16th another offensive sweep along the
west coast of Greece in the Zante area was more
productive, but a Beaufighter was lost. This
occurred when a mine-sweeper type of vessel of
100 to 150 tons was found in Zakinthos harbour.
The Beaus dived in using cannon and RPs, and
direct hits were seen on the vessel; but as they
turned away, Lieutenant Ridley and Lieutenant
Louw in 'N' reported their port engine coughing
oil and height down to 1000 ft. One of the other
Beaus was ordered to escort him, and Ridley told
to make for Taranto; due to the cloud, contact
was lost and 'N' failed to return.

January 1944 saw a number of changes taking
place in the Middle East, with the Allies having
numerical superiority over the *Luftwaffe* there,
and the movement of the ground forces up
through Italy, as well as the preparations for the
Second Front in the United Kingdom. 219
Squadron was one of the units that were returned
to Britain, and with the expected termination of
the war in Italy, there was expected to be a
movement of other units to the Far East for
reinforcement. Further to this the British had
plans to take the air war to the German forces in
the Balkans, by the allocation of certain
squadrons for this duty alone. Meanwhile,

squadrons would sweep the Aegean to cut supplies to the various outposts on the islands there.

One of the units that was engaged on this type of operation was 603 Squadron, carrying out convoy escort work and strikes on seaborne traffic and offensive patrols over the Aegean. During 27 January four of the squadron's Beaufighters were out on a sweep of the west and central Aegean, when they sighted west of Mykonos a formation of three Ju.52 floatplanes escorted by three or four Arado 196s. The formation was attacked and three Ju.52s and two of the Arados were shot down, with a third Arado 196 seen leaving the scene close to the sea; however it was not seen to crash. From the large amount of wreckage and burning on the sea, it was assumed that the transport aircraft could have been carrying fuel for the garrisons. As the Beaufighters headed home, LZ144 flown by Flight Sergeant Rooks and Flying Officer M. Thorne, suffered an engine failure and was forced to ditch, both crew members being seen to get into their dinghy.

On the 30th of the month, six of the squadron's aircraft, along with two of 47 Squadron Beaufighters, acted as an anti-flak escort to other 47 Squadron aircraft carrying torpedoes. The target was a convoy off Melos, which was found and attacked, the anti-flak Beaus going in first from almost sea-level with cannon and rockets. In spite of this, flak was intense, and hosed-up almost like a screen, which made assessment of the attack results impossible, but LZ272's crew claimed to have put three rockets into a 500-ton escort vessel. An Arado 196 acting as escort to the convoy got itself shot down during the proceedings, while as the Beaus turned away large red flashes and dark black smoke poured from a motor vessel.

Mosquitoes were by this date being fed into the Mediterranean theatre of operations for the re-equipment of Beaufighter squadrons — irrespective of the tendency of the Merlin engine's coolant temperature 'to go off the clock' on take-off. 108 Squadron based in Malta started to receive them in February and formed 'B' Flight of the unit. As regards the operating of aircraft in the North African Desert, it was estimated that the requirements for aircraft parts and replacements, were ten times greater than anywhere else. The Beaufighter and its Hercules engines — well, sleeve valves are noted for their hatred of grit — did not appear to be any worse off than any other piece of machinery. Airfields and ALGs became quagmires when it rained, and dust-devils appeared when it was hot; even in Italy, when the rains arrived, the airfields were seas of mud.

In February 39 Squadron had moved to Sardinia and were operating from Grottaglie, attacking shipping off southern France, firing rocket projectiles for the first time in March. 600 Squadron were now operating in Italy from Gaudo, and had been given authority to carry out long-range 'intruder' patrols, and were meeting up with night-flying Ju.87s! More and more Ju.87s began operating at night as the *Luftwaffe* withdrew its fighter aircraft for the battle over Germany.

On 9 February four Beaufighters of 252 Squadron walked into a trap, for during an offensive shipping strike in the east Aegean off Cape Zulufi, they sighted a small convoy escorted by two Ar.196s. As the Beaufighters attacked, three Me.109s dropped out of the sky on to them, resulting in one Beau going straight into the sea in flames. The pilot of one weaved and then climbed for cloud cover and escaped, seeing the other two Beaus heading south-west with the fighters chasing them; these two failed to return.

The occupation by the Germans of the islands of the Aegean, meant more targets in the way of ships for the Beaufighters; 252 Squadron on a shipping strike in the Leros and Arki area during February resulted in RP strikes on a six-ship convoy. On the 22nd 16 SAAF Squadron made a strike on Argastalion harbour, but quickly withdrew when intense flak made it impossible to sweep the harbour successfully.

Wing Commander B. Meharg took over as CO of 252 Squadron on 27 March, and the squadron combined shipping strikes with night intruder patrols during the month. Communications by air or sea to Rhodes and the islands of the Aegean received their attention. 16 SAAF Squadron were also operating over the same area carrying out shipping recce and shipping strikes, with the occasional 'beating-up' and shooting-up of any airfield that sheltered an enemy aircraft.

108 Squadron at Luqa was commanded by Wing Commander Banham, and were at this time

receiving Mosquitoes as well as operating their Beaus, with the intention of introducing night intruding to the squadron's night-fighter repertoire. For with the American landing at Anzio in January and the reduction of the *Luftwaffe* in Italy, 'trade' was now getting short for the night-fighters, who quite often only met night-flying Ju.87s!

Another night-fighting unit was 255 Squadron with their Beaufighter VIFs. In January 1944 they were based at Foggia, carrying out defensive patrols and escorts, particularly over the landing beaches at Anzio. On the night of 24/25 January Flying Officers Giles and Johnson on one such patrol were attacked from the rear by a Ju.88 night-fighter. This had made an unseen stealthy approach, and its first burst damaged the Beau's hydraulic system, elevator control and trimming tab, as well as liberally spraying the fuselage. After the attack Giles climbed the Beau to 10,000 ft, the EA having wisely departed, then returned to base and made a crash-landing without any injury to crew — apart from pride.

Later on during the 25th, Giles and Johnson were again airborne in a replacement aircraft, KV977. The sun was by now just below the horizon with little cloud and good visibility, when Johnson saw the silhouette of an aircraft at about four mile range. Turning the Beau towards the 'bogey', Giles soon identified the aircraft as a Do.217; this was carrying a bomb under its port wing and a smaller object under the starboard wing. On the approach of the Beau the EA took no evasive action, both aircraft travelling at 200 mph. Approaching from dead astern, Giles closed up to 300 ft and fired a short burst, with which the EA's starboard engine started smoking. Giles then gave the EA another burst at same range, at which the EA caught fire; a third burst blew pieces off the Do.217 and the main portion of the aircraft fell into the sea, Johnson thought that one person baled out.

On 22 April 252 Squadron sent twelve Beaus on a shipping strike, the target being the 2400-ton minelayer *Drache*, which was evacuating troops from Samos — the squadron ORB considered it a revenge for the previous setbacks at Cos. The attack was made from overland and was met by medium but inaccurate flak, but four 603 Squadron Beaus were acting as anti-flak and led the attack. The *Drache* was hit with about

five salvos of eight 60 lb RPs. Leaving the scene a Magda barge was cannoned and the *Drache* was now a smouldering mass with debris and fire rising into the air.

The same two squadrons were out again on the following day, seven 252 Squadron aircraft with five of 603 Squadron were again on a ship strike, the target was the SS *Orion*, which had been sighted in a bay of Denusa Island. Arriving at the target area the Beaus circled the island to make their attack from west to east, intense and accurate 20 mm and 40 mm flak immediately scarring the sky, hitting NV200 in the starboard wing and also damaging the aircraft's hydraulics. The *Orion* was repeatedly hit with cannon and RPs, but LZ456 was seen to be hit by tracer, jettisoned its RPs and broke away overland. NV200 force-landed at base and LZ456 failed to return. A further attack was airborne shortly after the first wave's return, but the vessel was found beached and burning, so more RPs were fired into it to prevent salvage.

During May 252 Squadron were called on to carry out intensive intruder programme and offensive sweeps over the eastern Aegean, mainly over every enemy landing field during the moon period. These intruder patrols had no success, but the offensive sweeps resulted in various smaller ships and lighters being taken apart.

June 1944 saw the formation of the Balkan Air Force, which was specifically created to provide support for the partisan forces in Greece, Albania and Yugoslavia, who were operating against the German occupation forces in their countries. One of the units selected was 39 Squadron, who moved to Italy with their Beaufighters for these operations. Two other units would also be included later on, these were 16 and 19 SAAF Squadrons. At this time the latter squadron was in the formative stage, being renumbered and converted from 227 Squadron by the posting-in of more SAAF personnel.

On 1 June a shipping strike was mounted that involved Beaufighters from 16 SAAF Squadron, 227, 252 and 603 Squadrons. The background for this was on the basis of intelligence work, for it was known that the German forces on Crete were getting short of supplies, and that in Greek harbours a number of ships were being loaded with supplies to form a convoy. This convoy included *Sabine*, *Gertrude* and *Tanias*.

The convoy was reported north of Crete by a 38 Squadron Wellington, so the Beaufighters, which had been gathered at Gambut, took off to attack led by Wing Commander Meharg and Flying Officer E. Thompson in NE293. 252 Squadron had been asked to contribute eight aircraft, but as ten were serviceable, these were sent. The convoy was located twenty-five miles north of Candia, and as well as the three principal ships, there were four destroyers and four smaller ships; the convoy travelling at approximately 8 knots.

On the approach to the convoy Meharg and Thompson in NE293 were hit by flak, and with one wing on fire dived into the sea. Flying Officers Jones and Wilson in JM233 of 227 Squadron were forced to ditch, but Flight Sergeants Sheldrick and Ash in JL897 shot down a Ar.196 and then attacked a Me.109, which dived away. 603 Squadron Beaus destroyed one Ar.196 and one Me.109 and claimed an Ar.196 as a probable, but lost LZ517.

The *Sabine* was hit by a salvo of eight RPs and sank; RP and cannon hits were obtained on the *Gertrude*, with direct hits on the superstructure, a huge sheet of flame shot up with black smoke, and although it reached Candia, it sank shortly afterwards. The destroyers and an armed trawler were hit, as was the *Tanias*; one of the destroyers sinking at Candia. Two motor vessels were severely damaged as was a UJ boat. So in spite of the air escort of Ar.196s and Me.109s, the Beaufighter had made a successful interception and execution on the convoy.

June would also see 108 Squadron saying goodbye to their Mosquitoes that they had received to supplement their Beaufighters; even worse was the news of the squadron's move from Malta to Edku — the squadron ORB stating: 'Squadron's disgust'. At Edku 108 Squadron were to take over the defence of the Nile Delta from 46 Squadron, while the latter squadron were to move to Tosca and St Jean for intruder operations over Cos, Rhodes and Crete. 47 Squadron with its TF.X Beaufighters had left the Middle East scene by this date and had moved to the Far East, under its CO Wing Commander W. Filson-Young. 600 Squadron had also moved, this time to Rosignano, and Squadron Leader Bailey had on one night shot down a Me.110 near Rome, followed shortly afterwards, on the same night, by a Ju.87.

An expensive patrol by 603 Squadron occurred on 23 July 1944, when six of the squadron's Beaufighters took off on a sweep of the central Aegean. First sighted was an Arado 196, which was shot down; then Flying Officer Bouneville and Sergeant Potter in NE494 reported damage to their aircraft and were unable to maintain height on its engines. This resulted in the aircraft ditching, but both crew members were seen to get into their dinghy. Four of the Beaus then sighted a convoy of one 800-ton corvette, one auxiliary sailing ship, one 300-ton coaster, two Danube barges and two smaller vessels. This convoy was attacked and the corvette and sailing ship claimed as damaged.

However, LZ340, crewed by Warrant Officer L. Sykes and Flight Sergeant Foxley, broke away from the attack with smoke pouring from its starboard engine, and were forced to ditch between Mykonos and Delos. NE610 also failed to return; it was later found that this aircraft had also ditched, and although the pilot was picked up by a convoy, the observer had drowned.

The intelligence control that the British had of the area is reflected in the continued story of Sykes and Foxley. Having ditched their aircraft, they carried out their ditching drill, then, possibly helped by a favourable wind or tide, they rowed to Delos. There they were taken over by a M19 agent, a message sent to CESME, following which a high-speed motor launch sped out and picked them up — thirty-two hours after they had ditched in enemy waters.

August 1944 would find 252 Squadron engaged on a search for a suspected U-boat in the Aegean; this resulted, for instance, on the 29th with eight Beaus flying in pairs carrying out sorties over the area from 0714 hours for the first take-off until 1710 hours for the last take-off — result, nil. On 6 September the squadron was on a shipping strike against a convoy sighted by a 459 Squadron Baltimore. The convoy was sighted south of Cape Sunion and consisted of a relief vessel and two escort vessels. The ship was identified as the SS *Carola* of 1350 tons, then in went the Beaufighters, attacking with cannon and 25 lb RPs. Many hits were seen on the motor vessel and the escorts, one escort being claimed as sunk.

A further strike by 252 Squadron was made on 19 September; Wing Commander Butler and

Flying Officer R. Crawford in NV373 leading eight Beaus to hit a 400-ton coaster in Gavrion harbour. One crew put four RPs into the stern, following which another four hits by the next crew on the stern resulted in a large explosion and the stern disintegrated.

Another victory for the RPs and the Beaufighters occurred in September, when on the 8th the Italian luxury liner 51,000-ton *Rex* had its ice-cream warmed up, when a number of squadrons, including No. 252 Squadron, sank it in Trieste harbour. 603 Squadron in October were sweeping the Dodecanese area, and on the 29th co-operated with FAA Hellcats in attacking gun positions on Melos. In January 1945 the squadron would receive orders to return to the UK.

As 1944 neared its end changes were being made in the re-equipment of a number of squadrons, and changes were planned with squadrons. 252 Squadron moved to Greece after being withdrawn from operations, the squadron providing air support for the Allied forces invading Piscopi, strafing targets indicated by a destroyer and taking out any resistance. The squadron's ground party disembarked at the port of Piraeus on 1 March 1945, and operating from

252 Squadron Beaufighters make a Victory Flypast over Athens 1945, with the Parthenon in the background. (Paddy Porter)

Hassani the squadron struck against Melos and Crete.

In late November 1944 39 Squadron had been taken off operations, and apart from six Beaufighters and crews, were to convert on to Marauders. The six-aircraft detachment going to Greece as part of the British contingent to suppress civil disobedience by Communist insurgents. 46 Squadron would reappear in the UK in a different role; with 600 Squadron being re-equipped with Mosquitoes, and 603 Squadron being disbanded in the UK. The new year was to find only three 'African' squadrons of Beaufighters still in being, for 255 Squadron were the next to receive Mosquitoes, with 108 Squadron being disbanded on 28 March 1945, and 272 Squadron being disbanded on 30 April.

The three remaining squadrons were 252 Squadron and the two SAAF units, 16 and 19 Squadrons, these latter two were operating from Biferno airfield against targets in Yugoslavia, attacking with cannon and RPs any enemy-held village, headquarters, barracks or buildings, strafing troops and taking out the railway system and trains — uneventful stuff, yet the final stages of German occupation.

It had been a long trail for the Beaufighters, from the UK to Malta, to Egypt, the long slog through North Africa to Tunisia, thence to Sicily and on to Italy. Strafing, bombing, night intruding and night fighting, the Beaufighter had

been there all the way; now the war in the Middle East had run its course — and the Beaufighter was still there.

Far East strikes

With the decision made to form a Beaufighter Wing in the Far East, the reforming of 27 Squadron at Amarda Road, India, was announced on 19 September 1942 by the Authority of Formation 287, which was issued by Air Headquarters India. So 27 Squadron was the first of three squadrons to be equipped with the Beaufighter and thus come under the control of 224 Group. Its first CO was Wing Commander H. Daish RAAF, who guided the squadron through its training period and on to operations, leaving the unit in July 1943. The squadron did not receive its aircraft until November, but found no problem in converting on to type, and the crews were more than pleased with their aircraft, which they considered magnificent and reliable, and potent enough to deal with any operation that they would be called up for.

The squadron flew its first operation on 25 December 1942 with a strafing attack on Taungoo airfield, but suffered so much trouble with the 20 mm cannon during this attack, that the squadron withdrew to Kanchrapara for two months to carry out rectification. The jamming of the cannon was eventually traced to defective ammunition, which was not helped by the humid and hot atmosphere. Trying to clear a jammed cannon in the Far East temperatures with the aircraft flying low and weaving, was not a recommended pastime.

A detachment under Squadron Leader Illingworth went to Agartala to carry out operations. Strafing attacks were made on airfields, troop concentrations, sampans, roads and railways, during which a Japanese Army 97 bomber was 'fired-up' at Taungoo, the squadron's first victory. Then the whole of the squadron moved to Agartala to carry out intensive operational flying over the months March–May. With the advance of the British ground forces, the Japanese would resort to the use of river craft of all descriptions to bring up stores and reinforcements — these then became the targets for the Beaufighters.

The greatest enemy of the Beaufighter, or any other low-flying aircraft, was the Japanese light ack-ack, which was very effective. The Japanese were also adept at laying trip-wires across the valleys to trap low-flying aircraft. These two factors alone were cause for anxiety, yet this does not appear to be reflected in the squadron reports; in fact, the opposite appears to be the case, for crews had confidence in their Beaufighters and its engines could absorb punishment. With their low, fast approach and relatively quiet Hercules engines, the Beaufighter was on target without warning — from the Japanese the Beaufighter received the name 'Whispering Death'.*

27 Squadron's range of targets was wide and the Beaufighter's twin engines and long range made it ideal for long-range sweeps into enemy-held territory, strafing railways, warehouses, river steamers and any military establishment. Attacks were also made on Heho airfield, Prome and Shwebo airfields. On 17 February 1943 two Beaufighters, EL344 and X8094 carried out an attack on Heho airfield, where some Japanese Army light bombers were attacked and set on fire, and a general strafing of the airfield carried out. EL344 flown by Pilot Officer Townsend and Pilot Officer Wandless failed to return, having last been seen during the first attack on Heho.

As has already been said, Japanese light ack-ack was very accurate, as would be demonstrated many times in Burma and in the South-West Pacific Area. Being demonstrated on 21 February, when Squadron Leader Illingsworth led six Beaufighters on an offensive sweep of Prome town and airfield. The attack went in successfully, but during the action T5274, flown by Illingsworth and Sergeant Osguthorpe, when attacking a target north of the town, was seen to be hit, and as the aircraft climbed pieces were seen to fly off it. Just as the pilot appeared to have steadied the aircraft, it did a steep climbing turn and dived into the ground.

On 1 March the Group HQ called the squadron to 'readiness' to meet possible requests from 4 Corp HQ for direct ground support; the squadron stood by in anticipation, and for the next two days, but no request came. Then on the 5th an attack by six Beaufighters was called for on the airfields in the Akyab area, and these were carried out successfully, except that EL295

---

* There appears to be some disagreement over this, whether it was British or Japanese inspired; but it was the Japanese who heard the whispering!

flown by Wing Commander Daish stopped a large bullet in the starboard undercarriage, that burst the tyre as well. During the subsequent night landing with the damaged undercarriage, the aircraft swung off the runway and the port undercarriage collapsed. Daish was unhurt and his observer received only a slight cut — any crash-landing you could walk away from is a good landing, and this frequently happened with the Beaufighter, so crews spoke proudly of the Beau's strong construction.

Nos. 176 and 177 Squadrons were formed after 27 Squadron, 176 Squadron coming into being at Dum-Dum in early January 1943, receiving its first aircraft on 13 January, when three second-hand Beaufighters were received from Egypt. The formation of the squadron commenced when a detachment of 89 Squadron arrived at Dum-Dum from the Middle East on 14 January; this was formed into the nucleus of 176 Squadron under the command of Wing Commander J. O'Neill, with a mixed bag of Beaufighter Mk. I and VI aircraft. The squadron was destined to provide the night-fighter defence of the Calcutta area, and received its 'blooding' on 15 January. At 2145 hours X8164 (Flying Officer Gray) and X7776 (Flt Sergeant Pring) scrambled on the approach of EA up the Hooghly River-mouth towards Calcutta. After a number of vectors by the GCI, Pring and Warrant Officer Phillips in X8161 got a contact on their AI, shortly afterwards a visual, completed by the shooting down of three Army 97 bombers.

Four nights later, X8164 (Flying Officer Crombie) and X8161 (Sergeant Fisher) were scrambled at 2100 hours on the approach of further enemy bombers towards Calcutta. GCI vectored them towards the EA, which were four Army 97 bombers, a visual was made and two more EA were going down as well as a probable. X8164 was set on fire by enemy action, so Crombie and observer Warrant Officer Moss baled out, returning to the unit overnight.

This loss of a large proportion of their bomber force in that area appears to have put the Japanese off, because although the occasional Japanese recce aircraft was seen, no more bombers approached. This resulted in 176 Squadron spending many hours on GCI practice, patrols, night-flying tests and other non-operational flying hours through the remainder of

January and February, without a sniff of the enemy. By March the squadron had moved to Baigachi airfield, yet still no activity by the enemy at night; but as the clear skies with a full moon approached in mid-March, squadron hopes were raised, as it was suspected that the enemy would take such an opportunity to bomb Calcutta; especially as information had been received that the Japanese had moved up heavy bombers to a forward base in Burma. Hopes were high, Beaufighters maintained at 'readiness', crews stood by — to no avail, the enemy did not come; maybe those two nights in January had taught them a lesson.

On 23 March Flying Officer Crombie and his observer were in the news again, but not in quite the same vein as past. Carrying out some firing practice in the mouth of the Hooghly, they then turned for base and were assailed by missiles from a RIN motor vessel. His were registered on the starboard engine and propeller, but the aircraft returned safely to base. The RIN vessel claimed that they had been attacked by an enemy aircraft and had hit it, as it was last seen going away with smoke pouring from the one engine.

With the lack of night intruders and only an occasional glance at a high-flying Japanese recce aircraft, it was decided to remove all unnecessary equipment from one of the Beaufighters and to carry out a flight test to check on the Beau's ceiling — the idea behind this was to try and beat the Hurricanes to the recce aircraft. The first attempt was made on 2 May 1943 by Flight Lieutenant Nottage and Warrant Officer A. Holton in V8503; however, as the Beau reached 21,000 ft the starboard engine air filter broke loose and so the attempt had to be called off. A further number of tests were made over the month, but only one reached 28,000 ft, most of the others failing through unserviceability of the aircraft or aircrew!

Around about this date a number of Hurricanes had been fed in to the squadron for interception purposes, but by the 12th of the month the squadron Beaus, going by the squadron ORB, were mainly on inspection, having spent so much time on night-flying tests, GCI practices and exercises, and gun calibration. This resulted in the squadron only being able to muster two serviceable aircraft for a four-plane 'readiness'. Serviceability was not improved either, when on the 20th X8022 during take-off, veered off the runway into a 90-degree swing, charged across a large irrigation ditch and finished up in a field. Although the aircrew were not hurt the aircraft was a total write-off.

In the Far East, by August 1943, there were four squadrons equipped with the Beaufighter, these were 27, 176, 177 and 211. The latter squadron had just re-formed and had both Blenheims and Beaufighters on its inventory. On re-forming it was based at Phaphamau, but after its armament training it was moved to Silchar, which was its base as a coastal strike squadron; there it was commanded by Wing Commander P. Meagher DFC.

On 4 August the command of 27 Squadron was taken over by Wing Commander J. Nicolson VC, still suffering from the burns he received during the Battle of Britain; yet even though still having to attend hospital for treatment he led the squadron on operations, a most worthy successor to Wing Commander Daish. Attacks were still being made on sampans and the Burmese railway system, and on the 17th a strike was made on an oil storage tank, which on being hit, opened out like a flower as it exploded.

September would find 27 Squadron completing 400 sorties since its debut on the Beaufighter, and from these sorties eight aircraft and crews were lost. Amongst the squadron's many claims for the period were sixty-six locomotives destroyed or damaged, and the damage or destruction of a large quantity of rolling stock and buildings. A typical example of this had occurred on 24 August, when two Beaus, EL298 and EL471, had cannoned a loco and carriages five miles from Thazi, then an attack on thirteen 3-ton lorries and two water lorries.

On 27 October two Beaus of 27 Squadron, SL590 crewed by Flying Officer Thorpe and Sergeant Chippendale, with SL767 crewed by Flight Sergeant Clegg and Sergeant Brinded, were on an offensive sweep of enemy river craft over an area Monywa-Pakokku-Thanbya-Tilin. The *Maha* was found and each aircraft poured bursts into it, photographs of the attack being taken. Then a steamer 115 ft long, suspected to have been the *Wuchango*, was located at Yesagyo, and nine strafing attacks made on it. The mention of a camera, may refer to the camera that 27 Squadron fitted to one of their Beaufighters, which originally had been fitted to

EL295 in January 1943, and was installed in the nose and used for head-on photographs.

During October 89 Squadron were withdrawn from operations in the Middle East and posted to the Far East, being based in Ceylon with Wing Commander W. David DFC, AFC as its CO. At this period both 27 and 177 Squadrons were heavily engaged in an offensive against all Japanese communications, whether road, rail or waterways; with attacks in the areas Gyodaung, Theddur, Thaki, Saigang and Kyonske. At 176 Squadron, Wing Commander H.G. Goddard was CO, and the squadron was based at Baigachi; unfortunately for the squadron it was still operating both Hurricanes and Beaufighters, with detachments pushed out to a number of airstrips.

This situation was to change towards the end of 1943, when 176 Squadron would become a full Beau squadron. The Japanese Army was now poised to push from Burma to India, trying to build up its reserves, in spite of the low-flying air attacks — yet how much reserve material *would* have been built-up if there had been no air attacks? At Imphal the Allied defences were being built up to resist this Japanese attack, but that would be in 1944.

In April 1943 27 Squadron received the first Mosquito for trials, and by the end of the year 'A' Flight had been equipped with Mosquito VIs, the squadron having thirteen on its strength. Although it has been said that these were for operational trials only, that is not the impression from the OR Book. Normally when aircraft were sent to a squadron for trials, the number involved did not normally cover the almost normal complement of the squadron. One gains the impression that neither the groundcrew nor aircrew were over-impressed with the Mosquito compared to the Beaufighter; certainly, the squadron were pleased to revert back to the Beaufighters, when by 9 March 1944 the Mosquitoes had departed.

Whereas in Europe the Mosquito was nicknamed the 'Wooden Wonder', in the Far East it was often referred to as 'Termite's Delight'; whilst this in no way reflects on the Mosquito's capabilities, especially in Europe, numerous crews in the Far East squadrons had less confidence in it than they had in the Beaufighter. One of the factors causing this attitude may have been its lower serviceability, as well as the Merlin's habit of boiling quickly prior to and during take-off in areas of high temperatures.*

One pilot of 27 Squadron who flew the Mosquito as well as the Mks. I, II, VI and X Beaufighters, said of the Mosquito: 'The Mosquitoes which appeared late in 1943, although much faster [than the Beaufighter], were not as tough and were prone to flak damage, especially in the radiators, the Japanese light ack-ack being highly accurate.'

Not only were the Japanese anti-aircraft guns very accurate, but the Japanese very adept at camouflage; so that even after strafing a column of MT and setting fire to it, a recce of the exact area shortly afterwards would find no sign of damage or fire. Bamboo and thatching were used to hide all transport, even locomotives, from the prying eyes of the RAF — this art of camouflage was of course helped by the normal foliage of the forests. The only exposure of the camouflage was by photographs, and this method was too late for the Beaufighter pilots during their low-level attacks.

The formation of 177 Squadron placed them under the control of 224 Group, and by 11 May

1943 four Beaufighters, JL712, EL349, EL352 and EL477, had been collected from 308 ASU Allahabad; the flight to base revealed rough operation of both the flaps and undercarriage, so all aircraft hydraulic systems were bled to rectify this situation. Then between the 24–28th of the month flying ceased due to leaking fuel tanks, during which period Wing Commander Baldwin assumed command of the squadron.

Further aircraft continued to arrive, but on 18 July during practice flying over the low-flying area, EL432 had its starboard engine falter, both propellers bit the water and flew off, the aircraft crashing on the river-bank with neither crew member injured.

Between 16–20 August the squadron moved to Chittagong, to be nearer to their area of operations. From there the squadron operated against all manner of Japanese communications, whether it was by road, rail or river; anything that moved in Japanese-held territory was strafed. Over the following months barracks, troops, trains, sampans, coaches, lorries, water

---

Result of a low-flying Beaufighter attack — things got in the way; trees, pagodas etc!

* R&M 2600 covers accident investigation, and indicates a fairly high structural failure rate for the Mosquito. It also details the strip investigation of a Mosquito returned from the Far East, separation and delamination of some parts having taken place.

towers and rail junctions were attacked continually, depriving the Japanese in the forward areas of supplies and reserves. Thus, on 22 September a strike was made typical of this period; two Beaufighters, JL536 flown by Flight Lieutenant Gandy and Flying Officer Buckley with EL519 flown by Flying Officer Bayard and Flying Officer Seary, attacked barracks and station buildings at Gyodaung, then shot up sampans and destroyed a locomotive.

Two days later two Beaufighters made a strike at Theddur, Thaki, to hit twelve coaches; but one was hit in the port wing and had to ditch in the sea off Chittagong; the crew survived. This was followed by two more Beaufighters making attacks on a 150-foot paddle steamer and another steamer; both were left on fire. In spite of intense ack-ack fire both aircraft returned safely. A little different to an attack on 30 September, when Wing Commander Baldwin and his observer, flying JL505, were forced to land in our territory fifteen miles south of Cox's Bazaar. Neither of the crew was injured and they were able to walk back with an Army patrol.

At the beginning of October 177 Squadron Beaus carried out attacks on two steamships, the SS *Maha* and SS *Assam*, then back to the mundane targets such as locomotives and troops. On the 10th JL937, flown by Flying Officer Bayard and Pilot Officer Seary, failed to return from an attack on such ground targets; the aircraft was last seen entering cloud.

Meanwhile, 252 Squadron in the Middle East had been detailed for movement to the Far East, but due to operations in the Aegean area a request had been made to delay this. This suspension of movement was agreed by the Chiefs of Staff, and would continue into January 1944. By May 1944 three further Beaufighter squadrons had equipped or arrived in the Far East war theatre: 22, 47 and 89 Squadrons. 47 Squadron had arrived in March equipped with Beau Xs, and had a vast experience of low-level strikes with cannon, rocket or torpedo.

22 Squadron had originally operated Beauforts in the United Kingdom, then moved to the Far East in 1942, and had been based at Ratmalana in Ceylon with the role of ship strike and coastal reconnaissance. Then in May 1944 they had commenced re-equipping with Beaufighters, but continued operating their Beauforts. On 14 May four of their Beaufighters took part in an attack exercise on Trincomalee; coming in fast and at 300 feet the attack was made before the defending fighters could make an interception. This exercise was repeated the following day for the training benefit of the defenders; this time the three Beaufighters came in at sea-level, pulled up to 100 ft above the ground and then zoomed to 3000 ft over the target, and again got away before being intercepted.

The commanding officer of 22 Squadron at this period was Wing Commander J.M. Lander DFC, RAAF, and the squadron pilots were reported as tending to fly the Beaufighter like the Beaufort, with the result that their target results in practice required improvement — which had been achieved by the end of July. By then the squadron had moved camp from Ratmalana to Vavunuya, settled in to their new abodes and liaised with a detachment of 47 Squadron based there.

The role of the squadron had then changed from the torpedo attack to use of rocket projectiles, with torpedo-bombing capability in reserve. The first Beaufighter was lost on 31 July, when during an 'overshoot' the starboard engine throttle control fractured and the aircraft stalled in and overturned. The crew escaped injury and there was fortunately no fire — the Beau again proving to be a tough old bird with strong protection for the crew.

Meanwhile 89 Squadron were flying Beaufighter VIs on night interception over Madras, with Wing Commander David handing over the squadron to Squadron Leader A. McGhie on 4 March 1944. On the same day the squadron was to lose X7961 and V8511 with engine troubles, which caused them to crash with the loss of four personnel.

In March 1944 176 Squadron were called on to provide a detachment at Imphal to support the British ground forces. The squadron's nocturnal activities however were not producing the 'victims' they should have done, partly due to shortage of 'trade'. For many hours were being flown on GCI practice and interception, the squadron having now become totally equipped with Beaufighters; the command had now passed to Wing Commander G.R. Nottage.

One pilot who prefers to remain anonymous, recalls one crash-landing on a Far East airstrip:

Apart from the crunch of bending metal, everything went quite straightforward for a

'Chocks away', 27 Squadron Beaufighter preparing to leave on a strike. Note D/F loop in plastic cover above cockpit.

belly-landing, though I suppose I could have made a longer approach; the direction started to change, and we slewed through a basha, over a bit of rough and into bushes. Strange to say, but I wasn't frightened because I was sure of the Beau's strength. Finished up with the Herc's each side bent a little, smashed up wings . . . but we were OK.

He felt sure that his aircraft was later repaired, as it was taken apart carefully — could have been because of a shortage of spares!

Beaufighters were more often damaged by the Japanese flak than Zero fighters, and as with most units in the Far East the damage was almost all repaired at the squadron, with erks working in the stinking heat and humidity. Fortunately the Beaus were Hercules powered, and thus could not only absorb punishment but also were better suited to the climate; the aircraft mainly being Mk. VI, X or XI, most with dihedral tailplanes.

A strike Wing for the Far East had been considered when 47 Squadron had arrived at Cholavarum, Ceylon, during April 1944. So that when SEAC was formed, 27 Squadron strike fighters were matched with the Torbeaus of 47

Squadron in 225 Group with the intention of forming an anti-shipping Strike Wing for use in the Bay of Bengal. Although 47 Squadron maintained its Beaufighter Mk. Xs at readiness during the month, no call was made on them. Whilst the supporting unit, 27 Squadron, after a period of training with 47 Squadron reverted back to its normal role.

April was a successful month for 177 Squadron, for during it they had carried out 165 sorties, which was the highest in any month since the squadron was formed. One outstanding feature was a period of ten days, when 177 Squadron in co-operation with 211 Squadron had made sustained attacks on long-range targets; these included the notorious Burma–Siam railway and East Burma as far as the Salween River. Unfortunately this was not carried out without loss of aircraft and crews, three being lost in one operation on 5 April.

On the 16th, one Beaufighter crewed by Sergeants Smith and Storey was severely hit by ack-ack and forced to ditch five miles out of enemy territory. In spite of a search they were not rescued for forty-four hours, when they were picked up by a Catalina escorted by 177 Squadron Beaus. Then on the 22nd, during an attack on a stores dump at Payingazu, the Beaus were intercepted by an 'Oscar', and although no combat was seen to take place, Flying Officers Gurski and Hacker failed to return.

During August 1944 176 Squadron had been transferred to Vavuyina, with a detachment at Ratmalana; then on the 18th a further detachment was sent to St Thomas Mount. The squadron became even more confused when its base was transferred to Minneriya on the 21st. Its task at this time being a mixture of night fighting and intruding, plus operating in the ASR role and co-operation with the Royal Navy — none of which was highly productive to a night-fighter squadron.

During August 177 Squadron were operating from Chiringa, and tasked to attack all Japanese communications, their CO now being Wing Commander J.E. Hill. During September the squadron was detailing pairs of aircraft on to certain communications; on the 9th two Beaus had the task of sweeping from the Bassein to the Irrawaddy, and two more swept the Bassein of all rivercraft. This resulted in two factories being strafed, a small steamer and a number of barges being damaged or destroyed, with Japanese troops made to scamper for cover.

The following day a further sweep was detailed, NE753 and LZ231 with crews sent out to strafe railway targets from Mandalay to Thazi, and to ascertain damage to a bridge at Myitnge. During the sweep three locomotives were damaged and rolling stock strafed, a water tower cannoned and some water craft damaged. As they moved on to Shwemyo two locomotives were seen in a shelter to the side of the track, earth blast walls being used now to protect waggons as well as locos. The locos were attacked and one appeared damaged as steam or smoke started to issue from the cover. The attack was then broken off as light machine-gun fire began weaving up at the Beaus; NE753, crewed by Pilot Officer D. Anderson and Warrant Officer R. White, was hit in the starboard engine oil cooler. Anderson feathered the propeller and proceeded to fly to Mytkinawa, passing this by R/t to the crew of LZ231, who escorted NE753 to a safe landing.

In October 47 Squadron were at Yelahanka, the unit equipping with Mosquito aircraft and the Beaufighters that had been flown to Nagpur. However, a number of Mosquito aircraft in the Far East had suffered structural failure, which resulted on 7 November with a SEAC signal being sent to Mosquito units that grounded all Mosquitoes for an inspection of the structure. 47 Squadron's aircraft were inspected, and that resulted in them all being grounded for structural faults as from 20 November.

176 Squadron Beaufighter fitted with AI.Mk IV radar at Digri, India. (Paddy Porter)

Ten days later the decision was made to re-allocate Beaufighter TF.X aircraft to 47 Squadron, and the squadron was moved to Ranchi for refresher training. This included low-level cross-country flights and 20 mm cannon ground firing. The squadron made its first flight with Beaufighters on 3 December, and upon completion of the training programme, they were, on 20 January 1945, moved back into the battle for Burma, being posted to Kumbhirgram; tasked to recce the roads south of Mandalay and to maintain a continuous night patrol over the Meiktila group of airfields.

In the meantime, 27 Squadron had handed in their Beaufighter Mk. VI aircraft and been issued with Mk. Xs — not exactly received joyfully by all aircrew, for although the Mk. X was higher powered with strengthened wings, it also weighed more and carried more equipment — worst of all it was slower! To be exact, about 20 mph slower. 27 Squadron also moved to Ranchi for training with the RP; this was in September, but by mid-October the squadron was returning to operations and moving back to Agartala in the squadron's original role. The CO of the squadron, Wing Commander Nicolson VC, having by this date been awarded a DFC for leading the squadron.

The 25 December found a crew of 176 Squadron being scrambled against a 'Bogey'; this turned out to be a 'Lily', and upon being sighted it dived to ground level from 7500 ft, carrying out evasive action. The Beau followed it down and poured in some cannon fire and saw strikes on the starboard engine and fuselage, then lost contact.

October 1944 saw 177 Squadron fly its 1000th operational sortie and it also established a claimed record of 200 locomotives damaged or destroyed. However, on the 5th the squadron lost is CO, Wing Commander J. Hill and his observer Flying Officer Broughton, when their Beau NE754 failed to return from a sortie against road communications in the Namsung-Lashio area; Hill was a highly respected CO of the 'press-on' type. The squadron, along with the other Beaufighter squadrons, was amongst a strike force of approximately 200 Allied aircraft that attacked Mingaladon airfield near Rangoon on 18 October.

October found 89 Squadron operating from Baigachi, with a detachment sent to Tulihal, to operate over Imphal. On the 26–27th Pilot Officer Vignall and Flying Officer J. Ashworth crewing X8141 were detailed to attack Heho airfield. Taking off at 2130 hours the crew set course for Heho, and upon arriving there they commenced to strafe the aircraft pens and store buildings. On the return flight it was found that at 0215 hours the fuel was running low, so the aircraft was climbed in case of a need to bale out. Then on the airfield circuit the one engine cut out, with the other engine cutting out as the aircraft touched down!

November was mainly spent on night 'Rhubarbs' strafing ground targets, although day operations were concerned with strafing vehicles; the squadron operating from Tulihal and Chittagong. Some priority was given to attacking the railways in Mandalay from Thazi to Saigiang. On the 21st Wing Commander F. Collingridge took over as CO from Wing Commander McGhie.

On 20 December a 'bogey' was notified to 89 Squadron, so Squadron Leader R. Morrison and Warrant Officer Powell in V8710 were scrambled. A contact was made, followed by a visual, which was a four-engined aircraft; Morrison called up the controller with this information. After repeated requests to the controller for confirmation that it was an EA, confirmation was given, so Morrison closed in to 700 ft and opened fire, continued to close to 500 ft. When the EA burst into flames Morrison recognized it as a Boeing B-29 — there were six survivors.*

22 Squadron was by this date at Kumbhirgram in Assam and under the control of 221 Group; due to a shortage of ground personnel the aircrew were having to help with the aircraft daily inspections and compass swinging. Then on 23 December NV328 was badly damaged on take-off, when the pilot, Squadron Leader R. Gee, had trouble with the port engine as the aircraft accelerated, so decided to abort and cut back the throttles. Due to its momentum the Beau continued to roll slowly forward to the end of the airstrip, but most unfortunately and unknown to the pilot there was a large hidden ditch at the end

---

* The Squadron ORB states a B-29, and 22 Squadron ORB dated 11 August 1944 also mentions 'Operation Boomerang' and B-29s, but the author is in some doubt as whether these are recorded errors for B-17; the B-17 was the aircraft operating over this part of the war theatre.

Mk.VI based on a strip near Mandalay. Aircraft fitted with AI.Mk IV radar.

of the runway; this resulted in the wheels entering the ditch and the aircraft turned over on its back — neither crew member was seriously hurt.

On 11 January 1945 fourteen Beaus of 22 Squadron had moved to Dohazari to operate 'Rhubarbs' under the control of 901 Wing; first of all they carried out sector patrols for familiarization of the area. Then on the 18th six of their Mk. Xs flew their first sorties in the Irrawaddy area from Henzada to Chauk, river craft were strafed and a large oil barge blew up upon being hit. As opposed to other squadrons' low-level operating height, 22 Squadron flew its sorties higher at 500 to 1500 ft, as it was felt that this gave a better vision on the approach to a target.

Two days later Wing Commander Lander lost his life in an accident to his jeep, when returning from a visit to another unit, so Squadron Leader Gee assumed temporary command, later taking over as CO and being promoted Wing Commander.

In November 1944 27 Squadron lost two Beaus during a low-level strike over Burma, when a formation of Lockheed Lightnings flown by 'green' USAAF pilots bounced them, shooting down both aircraft and killing the crews — the aircraft identification capability of USAAF pilots must be suspect, as the same thing was happening in other theatres of war as well.

Then in early 1945 the squadron's operational strength was reduced through lack of spares and the need for the airframes to be overhauled; the main casualty being tyres, which were taking a beating due to the condition of the forward airstrips.

The squadron had, by the end of January 1945, flown 160 sorties in the month and lost only one aircraft; pressure was by now decreasing as the Japanese withdrew eastwards, which resulted in some of the crews being detached to Armament Practice Camp. Then on 20 March Wing Commander T. Bradley was posted in as CO of the squadron, and rumours abounded of a change of role for the squadron — squadron life would never exist without the circulation of good rumours!

January 1945 found 47 Squadron in action; on the 26th Wing Commander Filson-Young led four Beaus on a strike on a building on the Mandalay waterfront, which was suspected of being an enemy supply centre. Direct hits were made on the building by all thirty-two RPs, and Japanese soldiers were seen hightailing it to the horizon. This was followed on the next day with nine of the squadron's Beaus carrying out a dusk to dawn patrol over a concentration of enemy artillery, which was threatening the 19th

Division bridgehead on the east bank of the Irrawaddy, forty miles north of Mandalay. Bomber and fighter bombers had made a concentrated attack during daylight, and the presence of the Beaus ensured a peaceful night for 19th Division.

On 9 January LZ533 of 177 Squadron, crewed by Flying Officers MacIntosh and Royle, was returning from an uneventful patrol, when they saw a light flashing, this read: 'SOS, rations 400 men Recce Regiment'. Circling around MacIntosh replied, 'Maintain smoke signals, am in touch with HQ. Wait.' MacIntosh then contacted 10 Operations, who asked him to land at Cox's Bazaar. HQ then took over.

Due to the constant attacks on Japanese communications, targets were getting harder to find, with the result, that to extend the range of the Beaufighter, attention was being paid on how to increase the air miles per gallon of fuel; this was finally written up into a readable form, and started off thus:

> Since the constant attack on Japanese lines of communication in Burma and Siam began to take effect, it has become increasingly difficult to find suitable targets for Beaufighters, and as a consequence much attention has been paid to extending the range, and therefore the sphere, over which these aircraft can operate . . .

RD713 Beaufighter Mk X fitted with AI. Mk VIII, seen at Karachi, India.

The 7 February found 47 Squadron detailed to patrol over 19th Division's bridgehead at Kyaukmyaung, forty-five miles north of Mandalay; the first aircraft on patrol found no artillery activity, but managed to shoot up and damage eight MT vehicles. Unfortunately the third aircraft failed to return and the fourth crashed in flames five miles from Imphal. Then on the 11th the squadron carried out the first 'Cloak' operation; this was to simulate an attack in support of a 14th Army crossing of the Irrawaddy, by dropping of 'Pintails' (automatic-firing Very cartridges which fired on impact with the ground). One Beau was detailed to drop six 'Pintails' in each of three areas, timed to be between Mosquitoes dropping 'Farafax' (dummy artillery fire) and Dakotas dropping dummy paratroops.

On 26 April 1945 177 Squadron were to lose their CO, when Wing Commander G. Nottage and Pilot Officer Bolitho took off in RD376 to attack and possibly photograph a probable Japanese radar site. Due to bad weather in the target area with 10/10 cloud down to ground level, it was decided to attack targets of opportunity. The first attack was made on a lugger loaded with Japanese troops and cargo, then two more luggers and an escort motor launch were sighted. The escort vessel was attacked with the remaining ammunition and Nottage turned away. Back at the squadron none of this was known, as no messages had been received, so the fateful words FTR would be written.

On 6 May Nottage and Bolitho made contact with a special army field force known as 'E' Group 136 Special Force Party; then the story was told. As RD376 turned away Nottage believed the aircraft had been hit, the starboard engine oil pressure dropped, but both engines kept running; so he climbed the Beau to 2000 ft for safety, when suddenly the starboard engine cut with a loud bang and the propeller would not feather, the aircraft shuddering. The pilot headed NE but the aircraft would not maintain height, and he was forced to land, coming to rest in a paddy field. Nottage, who had only just got over fever, was still very weak, nevertheless they started their march back, and although the native population saw them no help was given until a few days into their march, when a village headman and others gave them food and information.

After their perilous march through enemy territory, nerves atingle, short of food, they could not find praise enough for these special army forces that constantly operated in enemy environments. Their escape was signalled back to HQ and a clearing made, then in came a Jungle Rescue light plane to whip them back to civilization.

By this date 27 Squadron had its role changed to Air Jungle Search and Rescue, and split into three detachments, one at its base at Chiringa, four crews and three aircraft at Akyab, and five crews and three aircraft at Monywa; the Monywa detachment being later moved to Meiktila. Their Beaufighters carried emergency containers Mk. Is from their torpedo slings, and in this role the squadron were to continue until VJ-Day.

By March 1945 the war in Europe was grinding to its end, the war in the Far East was seeing the re-conquest of Burma, and the Beaufighter squadrons in the Far East were declining in numbers, as the Air Ministry in the UK were standardizing aircraft types irrespective of their suitability to area and climate, with the Mosquito being one of the standard types — this would result in 1946 with stranded unserviceable Mosquitoes at numerous airfields in the Far East. 47 and 89 Squadrons had, during February 1945, been withdrawn from operations and started converting to Mosquitoes, with 211 Squadron converting to the Mosquito in May.

176 and 177 Squadrons would continue flying Beaufighters until July 1945, when 176

Squadron began conversion to Mosquitoes, with 177 Squadron being disbanded on 5 July. 22 Squadron kept flying Beaufighters until 30 September, when it was disbanded. The Beaufighter had continued until the end — when the Japanese who heard 'Whispering Death' never knew what had hit them. She was a tough lady was the Beau, even when flying under the most primitive conditions and in probably the worse climatic conditions in the Far East — if not the world.

Code names for some Japanese aircraft encountered in Burma and the Pacific and Australasia areas.

'Betty'	Mitsubishi Navy Type 1 land attack
'Dinah'	Mitsubishi Army Type 100 recce
'Lily'	Kawasaki Army Type 99 light bomber.Ki48
'Nick'	Kawasaki Army Type 2 fighter. Ki45
'Oscar'	Nakajima Army Type 1 fighter
'Pete'	Mitsubishi & Sasebo Navy Type O observation seaplane
'Rufe'	Nakajima Navy Type 2 fighter seaplane. A6
'Sally'	Mitsubishi Army Type 97 heavy bomber
'Zeke'	Mitsubishi Navy Type O carrier type fighter

Action in the Antipodes

One only has to realize what stalwart, tenacious fighters RAAF and RNZAF personnel were in the defence of Europe, to understand their outstanding defence of their homelands. With their hands on such a magnificent strike fighter as the Beaufighter, suited to the climate as well as reliable, the aircrew were not satisfied with defence; wherever there were Japanese the RAAF Beaufighters struck — and struck again and again. Every Japanese ship, ground installation or troop concentration, on the sea or on any island within the range of the Beaufighter squadrons, would get a taste of Whispering Death.

The first RAAF unit to receive the Beaufighter was 30 Squadron, service commencing in June 1942, when the squadron commenced familiarization on their Beaufighters. These were British-built aircraft, that had been released in May 1941 after the British Government embargo had been revised. Authority by the Australian Government allowed the purchase of fifty-four aircraft, and these were delivered in RAF day camouflage. The first two released to the RAAF were T4926 and T4927 — built by Fairey Aviation — and were taken on RAAF charge on 26 March 1942. A further increase in orders for

Beaufighters for the RAAF took the total to eighty-seven by the end of 1942, with an ultimate purchase of 218 British-built aircraft. These were all delivered with British equipment and camouflage, but were re-serialized in the Australian A19 series upon arrival.

The RAF top surface camouflage was left as imported, until wear took its toll, when dark green and dark brown were applied, sometimes dark foliage green being used instead of dark green. The undersurfaces were all re-painted SKY 'S', and this was maintained whether the aircraft was imported or Australian-built. 30 Squadron's code letters were LY, but it was well into 1943 before the squadron started using them, so some photographs of 30 Squadron aircraft can be seen without any code.

30 Squadron's first official commanding officer was Wing Commander B. Walker, who took over the unit on 4 June. After a training period — brief because of the war theatre's proximity — the squadron provided escorts for some Beaufort strikes, before moving to their

operational base near Port Moresby. Their aircraft were ex-RAF Mk. Ic's and so the engine performance was rated at a maximum for low altitude, which suited the operations that the RAAF carried out. In flat-out performance, 'pouring on the coals' could soon keep a Jap fighter at bay, and was the best defence against the nippy Nipponese fighters. Even though it was a natural Australian action to 'mix it', experience soon showed that it was best to get in low and fast, strike hard and fast and exit fast.

By the time of the Beaufighter's entry into RAAF service, the Japanese enemy was eating its way through the Philippines to the Islands of the Dutch East Indies, and was threatening the security of the Australian continent. Darwin had been bombed on 19 February 1942; ships in the harbour were set on fire or sunk, the town and airfield bombed and twenty-three Allied aircraft destroyed. The Allied course lay in a holding action while sufficient Allied forces built up. For the RAAF, this meant striking at the enemy with anything all the time, so with the entry of the Beaufighter into service they had the means to strike hard, fast and low.

During early 1942 it became necessary to delete the red centre from the RAAF roundels,

'So you want me to fly close escort?' DAP test pilot Harold Skelton flying close off Black Rock beach. (Keith Meggs)

due to attacks being made by USA aircraft not recognizing the Australian markings! Then during the early flying of the Australian-built Beaufighters, the failure to fit tailwheel locks resulted in an unnecessary number of swings on landing, with expensive results.

In August 30 Squadron was based at Bhole River, Queensland, and there were twenty-seven Beaufighters in the inventory of the RAAF, but it was not until 7 September that they commenced operations. Beauforts of 100 Squadron were, along with three Beaufighters of 30 Squadron, to attack an enemy cruiser and destroyer near Normanby Island. One Beaufighter crashed on take-off, but the other two strafed the ships whilst the Beauforts carried out their torpedo attacks; neither ship was seriously damaged and withdrew.

Port Moresby was the key position in New Guinea for the Allied forces, followed by Milne Bay; so reinforcements were necessary to stiffen up the area. This resulted in 30 Squadron moving to Port Moresby on 12 September, and five days later putting in an attack on enemy barges at Sanananda Point and troops along Buna Beach. This was followed on the 23rd with a further attack at Buna, but this time light ack-ack brought down A19-1. The same day, a follow-up

attack by six Beaufighters got intercepted by 'Zeros', but after a first strike and a quick exit, the Beaus dived to sea-level at their favourite sprint pace and easily outdistanced the 'Zeros'.

During the next month or so the 30 Squadron Beaus would be involved in ship recce and strike, as well as strafing enemy rear areas in support of Australian ground forces, so denying the enemy his supplies and reinforcements, sometimes with the loss of aircraft to ack-ack fire, the weather or terrain. Work and flying were carried out in the most primitive and atrocious conditions, where the enemy of flying was not only the Japanese, but the climate, terrain and sea crossing as well.

Then on 25 November 1942 five Japanese destroyers were located in Huon Bay; so 9 Operational Group set up an attack, which commenced with a strike by Beauforts, followed by Hudsons bombing the enemy. After this came the sighting on 2 December of four Japanese destroyers off Buna; so at night a Hudson illuminated the target area with flares, whilst six

Ex-RAF Beau' in RAAF service in the bush, still with original roundels. (Frank Smith)

30 Squadron Beaufighter strafed the destroyers, to provide an anti-flak escort for six Beauforts going in on torpedo runs, one scored a direct hit and possible damage.

An enemy convoy was sighted on 6 January 1943 proceeding from Rabaul to Lae, the Japanese becoming desperate to reinforce with more troops and provisions, most of which would be destroyed by air attacks over a few weeks. An attack was made by a Catalina on the convoy, and on the following day Beaufighters of 30 Squadron could find only wreckage. Fifteen 'Zeros' attacked the Beaufighters and in the following combat three were shot down without loss to 30 Squadron.

The following day the Beaufighters joined up with USAAF Lightnings in providing cover to a bombing raid on Lae; this resulted in wrecking aircraft on the ground, an ammunition dump blown up, barges set on fire, and the sinking of the *Myoko Maru* of 4000 tons. This was followed by a further raid, after which the enemy convoy left port back to Rabaul, once again to be harried on its way.

About this time the Australian Army 'Kanga Force' were being driven back to Wau by the Japanese, 'Kanga Force' having harried the Japanese in Mubo, Salamauma and Lae areas with guerilla tactics. So reinforcements of troops were flown into Wau by C-47s, followed on 30 January with 25-pounder artillery pieces being

flown in. Beaufighters of 30 Squadron continued strafing the enemy, including blowing up an ammunition dump; so that in early February though the Japanese were still trying to take Wau, the Australian infantry and artillery backed by Beaufighters, Havocs and Mitchells were holding out. On the 26th of the month it was over and the Japanese were on the retreat from Wau to Mubo, prodded along by the Australian ground forces and Beaufighter strikes.

The second Beaufighter unit was 31 Squadron; this officially came into existence at 5 Aircraft Depot, Forest Hill on 14 August 1942, but its aircraft did not start arriving until the 20th. The squadron next moved to an airfield south of Darwin, before in October moving to Coomalie. It was from that airfield that six 31 Squadron Beaufighters commenced their operations, when on 17 November they attacked Timor; losing their first aircraft A19-46, flown by Squadron Leader Riding and Warrant Officer R. Clarke.

Throughout December the squadron became quite active, with the strafing of Japanese troops, stores, barges and buildings, sinking a sailing ship off Timor and shooting down a Japanese fighter. January 1943 was also to prove interesting, for on the 27th, based on information received from 'coast watchers', the squadron moved forward to Drysdale for a strike the following day. Taking off the strike attacked Penfui; going low they hit the airfield where Japanese bombers had arrived, destroying twelve bombers on the ground and damaging ten others. The Beaufighters were intercepted by three

'Zeros', but apart from two being damaged the squadron arrived back safely.

Meanwhile, a limited Allied offensive was in preparation in New Guinea, and the Japanese were likewise preparing new airfields in the islands north-west of Australia, to strike at the Allied base. For Australia was the only base territory upon which forces could be formed and supplied from. This resulted in the first few months of the year in an almost daily Allied air offensive, Beaufighters and other aircraft carrying out raids on enemy bases and shipping. For the Beaufighters in the SWPA (South-West Pacific Area), it quite often meant flying long-distance low-sorties on shipping strikes, picking out targets on the high seas or in between the islands. Although Japanese fighter opposition was met, as in Burma it was the light ack-ack which was the most effective. So that once a Beaufighter was partly disabled, it was easily dealt with by a Japanese fighter — and the Japanese treatment of prisoners of war was no consolation. Even after the Japanese surrender, a number of RAAF prisoners were executed.

On 2 May after an air raid by the Japanese on Darwin, four Beaufighters of 31 Squadron were scrambled to track the enemy back to their base; A19-16 was forced to turn back due to unserviceability, but the other three pressed on. Arriving at Penfui, the three dropped to ground level and swept in line abreast with all guns firing, catching the enemy aircraft refuelling and rearming, cannoning the airfield ground defences and aircraft. As the Beaufighters departed, zipping away out to sea, they left behind two destroyed 'Zekes' and two bombers, with smoke rising above the treetops.

This raid was followed by a further one on the 6th, with a raid on a number of Japanese bases, during which 31 Squadron Beaus accounted for nine enemy floatplanes for the loss of one aircraft, A19-60. The squadron was now operating from Millingimbi on a temporary basis, the airfield being frequently raided by the Japanese, during one raid on the 10th A19-72 was lost and three others damaged on the ground.

On 19 May 31 Squadron returned the compliment by attacking Penfui; although two bombers were destroyed on the ground, the Japanese ack-ack as usual was intense and accurate, which resulted in A19-28 and A19-29 being shot down and A19-45 being forced to ditch on the way back.

With the destruction of the convoy in the Bismarck Sea in March, the Japanese were forced to bring in reinforcements and supplies by any means possible. 30 Squadron were at Port Moresby at this date and on 9 May sent eight of

Remains of IC A19-55 of 30 Squadron, which was destroyed by enemy action on Wards Strip, Port Moresby, 27 January, 1943. (Frank Smith)

Groundcrew viewing damage to 30 Squadron A19-54.
(Frank Smith)

their Beaus to attack Madang; this was carried out without loss. Unfortunately, this was not always the case, for the enemy ack-ack was as accurate as ever.

So as to be able to make attacks on the enemy in New Britain, Goodenough Island was prepared as an advanced operating base, which resulted in two airfields being prepared near Vivigani. The first use of the base was by the Beauforts of 100 Squadron. Enemy barge traffic was active both by day and night and the Beaus and Beauforts sought them out. This was to continue as the Australian forces attacked along the Huon Gulf coast, and on towards Lae and Salamauma.

During May 31 Squadron were to carry out ninety-nine sorties and claim the destruction up to the end of the month of twenty EA. They then began June by raiding the Taberfane seaplane base on the 4th. This raid was made from Millingimbi, and the four Beaus swept the anchorage and shot down a floatplane in the process. The squadron was to return to Taberfane on the 12th, and again were to stage through Millingimbi. This time they swept in at tree-top height and caught the enemy napping, destroying seven floatplanes and damaging two. A further strike was made on the 22nd, but this time bad weather hampered the attack and the squadron lost two Beaus (A19-62 and A19-113) which crash-landed back at Millingimbi.

Goodenough Island was to be the base for a strike on 22 July against Gasmata airfield. This was a major attack of sixty-two RAAF aircraft, which included eight Beaufighters of 30 Squadron. Two Kittyhawk squadrons (76 and 79) flew cover over the base during the departure and return of the raid formation. The role of the Beaufighters was to strafe Gasmata airfield area; the raid was accomplished without loss. The raid was repeated four days later, again without loss.

The next stage in the campaign on the ground was the seizure of the Markham Valley, Lae and Salamaua, with the 7th Australian Division attacking along the Huon Gulf coast towards Lae. This was to be helped by the establishment of an airfield at Kiriwina, whilst 22 and 30 Squadrons were to operate from Vivigani and maintain a continual attack on the enemy barge traffic along the coast. This was to continue through the August to September.

During the start of September a number of Beaufighters were lost doing recces and strafing attacks; one of these was 30 Squadron's CO, Wing Commander C. Glasscock. He and his observer were killed when their A19-133 was shot down during an attack on Cape Hoskins. By this date Lae and Salamauma had fallen to the Australians, who were now on the move to Finschhafen. The main attack by the Allied air forces would now be the main Japanese base of Rabaul — guaranteed to give an airman the hottest reception in SWPA!

In spite of continual attacks on Taberfane seaplane base, the Japanese seaplanes from there repeatedly harassed Allied shipping, and on 10 August sank the MV *Macumba* of 2526 tons, despite protection being provided by three Beaufighters of 31 Squadron and a warship. Five days later nine 31 Squadron Beaus attacked Taberfane, but were attacked by ten 'Pete' seaplanes and hampered by the weather. Returning to Taberfane on the 17th the squadron Beaus were more successful, strafing small boats, destroying a lugger, and shooting down one 'Peter' and three 'Rufes'.

During October, 30 Squadron was to form part of a number of Allied attacks around Rabaul, on one occasion being attacked by USAAF fighters. Then on the 15th A19-97 was lost as well as a number damaged; this occurred when Allied aircraft began a total suppression of Rabaul, 308 aircraft taking part. Twelve of the squadron's Beaus were detailed for the operation but were delayed by the weather, and so reached Tobura

without a fighter escort. They were intercepted by one 'Hap' and eighteen 'Zekes', and during the combat two 'Zekes' were destroyed and the Beaufighter flown by Flying Officer Stone and Flying Officer E. Morris-Hadwell failed to return.

During November the squadron moved from Goodenough Island to Kiriwina, with strikes being made around Rabaul, Bougainville and Arawe. At this period the Japanese air strength at Rabaul was at its highest for the war, with at least 390 being fighters, so the vicinity of Rabaul was no health resort for Allied aircraft. On 15 December Allied forces had landed at Arawe, ground forces in Bougainville had completed the construction of an airfield and fighter sweeps of SWPA aircraft were attacking Rabaul.

The 13 November saw the formation of 10 Operational Group, which consisted of 77 and 78 Wings, neither of which included a Beaufighter squadron, but this would be rectified in 1944. October and November found the weather deteriorating with the onslaught of the monsoon, and thus the restriction of flying operations. This also applied to the Japanese as well, and only one raid was made by Japanese bombers on Darwin.

December found 31 Squadron making attacks on Japanese shipping, one on the 15th was quite productive — from the RAAF angle. Eight Beaufighters attacked Manatuto and sank two

Remains of A19-55 of 30 Squadron destroyed by enemy action on Wards Strip, Port Moresby, 27 January, 1943. (Frank Smith)

barges and damaged six schooners; they then sought out a convoy and sank a vessel of 500 tons. The same convoy was then attacked in the afternoon by five Mitchells, which sank the 5123-ton *Wakatsu Maru*. Early the following morning the Beaufighters were out searching for the remains of the convoy, which they attacked. Barges unloading the ships were strafed and sunk, whilst the cargo ship *Gennei Maru* was sunk at its moorings. Four 'Nicks' that intercepted the Beaufighter attack caused no damage and lost one of their number to the Beaus.

January 1944 would find the Japanese strength on the wane and the RAAF squadrons continuing to support the invasion of New Britain by attacks on the bases there and in New Ireland, as well as the traffic in between bringing supplies and reinforcements. On 13 February orders were issued for invasions, to the SPA forces to attack Kavieng on 1 April and for the SWPA forces to attack the Admiralty Islands on the same day. By this date Rabaul had been isolated, and had cost the Japanese the loss of approximately 820 aircraft and a large amount of shipping.

In March 1944 the Vultee Vengeances of the RAAF were withdrawn from 10 Group under orders from MacArthur, and 30 Squadron Beaufighters and 22 Squadron Bostons took their

RAAF Beaufighter in flight over New Guinea.
(Frank Smith)

place. The Beaus flying both day and night sorties under its CO, Squadron Leader F. Maquire DFC, in support of the various Allied forces in the SWPA, often resulted in flak damage to the aircraft and injuries to the crews; whilst maintaining the aircraft on forward airstrips tested the ingenuity of the groundcrews working under the most primitive conditions.

Operating from Coomalie on 28 March 1944, and staging through Drysdale, six Beaufighters of 31 Squadron attacked Roti Island, south-west of Timor. Striking at Pupela Bay in a low-level strafing attack, strikes were made on five sailing ships. However, A19-182 going in to the attack, was so low that it hit a mast, whipped into the sea and exploded. This attack was followed on the 31st with an attack on Tenau Harbour in Timor; this time cannon was supplemented with 250-lb bombs, and resulted in an oil barge blowing up and a 500-ton vessel left on fire; all the Beaus streaked away safely.

April started off with an attack on Semau Island, a low-level strafing attack cut down troops, set an oil barge on fire and destroyed a number of buildings. Light flak was hosing up as usual and hit A19-156, which eventually crash-landed on Cartier Island; the crew were safe and a rescue was made by a Catalina later. On the 15th another Beaufighter was lost, when, during an attack on Japanese installation in Su village, sustained and heavy ack-ack resulted in the loss

of A19-178. The Beau was seen to take a hit and crashed into the sea south of Timor.

General MacArthur had by this date issued orders for the invasion of Hollandia and Aitape, with Aitape being occupied first, so as to allow the preparation of airfields for the attack on Hollandia. In preparation for this, 78 Wing ground staff moved forward by ship; the invasion commenced on 22 April and the airstrips were ready for use on the 24th. Mining operations were carried out off the Caroline Islands and fighter patrols were over Hollandia and Aitape.

The 18 April attacks were made to help in the attack on Hollandia and Aitape; Spitfires of 54 (RAF), 452 and 457 Squadrons led by Beaufighters of 31 Squadron, commanded by Wing Commander Mann, staged through Bathurst Island strip and carried out attacks on Wetan Island and Barbar Island, all the aircraft returning safely. The following day, eight Beaus of 31 Squadron with fifteen Beauforts of 1 Squadron and twelve Mitchells of 18 Squadron, carried out an attack on the Su town area and barracks. With the assault on Hollandia on the 22nd, Beaus and Mitchells raided Dili, whilst Liberators bombed Noemfoor Island.

During May, targets on Timor were constantly attacked, and by the end of June the RAAF were supporting an amphibious task force attacking

A19-11 damaged as a result of bomb damage at Wards Strip, Port Moresby, 12 April, 1943. (Frank Smith)

Noemfoor Island. The Allies were now getting into their stride and MacArthur's spearheads were pointing towards the Philippine Islands. United States troops landed on Biak on the 27th, whilst in New Britain the Australian troops had cut off Wewak and its garrison, estimated as 30,000 troops. As Wewak was still considered a threat, even though isolated, 10 Group began attacks to reduce supplies, troops and their defences.

On 13 July 30 Squadron were to lose three aircraft and their crews during strafing attacks on barges, which were being used to support and re-supply the garrison in the Wewak area; these were A19-146, A19-174 and A19-185. The crews were experienced airmen that the RAAF could ill afford to lose, one of them being Squadron Leader G.K. Fenton. 31 Squadron were also busy during July, their Beaus joining up with 2 and 18 Squadrons in attacks on enemy barge traffic between Timor Babar, Sermata and Leti Island, and had on the 16th attacked Maumere airfield. During this latter attack the Beaus had destroyed three 'Nicks' before strafing shipping in the harbour, streaking away with columns of smoke rising from the harbour -

Hybrid Beaufighter of A19-205 tail and A19-142 fuselage made by personnel of 11 Repair & Salvage Unit, Noemfoor Island. (Frank Smith)

those that were hit never heard the Whisper of Death!

The Allies' 'island hopping' campaign was by now rolling forward, many strongly-defended bases being bypassed in preference for the lightly defended ones; the former were pinched out as they ran short of supplies. This process also allowed the building of airstrips for air strikes to move forward and reduce other garrisons. The 30 July saw the last amphibious operation in New Guinea, when Sansapor, Middleburg Island and Amsterdam Island were invaded. 30 Squadron moved forward to Noemfoor, which had been taken earlier on in the month, and continued their strikes and strafing; 30 Squadron at this time being commanded by Wing Commander J.T. Sandford, an officer who was on his second tour of operations.

September would find the RAAF squadrons fully occupied attacking; more Spitfires had joined squadrons, and by then the Australian-built Beaufighters were starting to reach the squadrons. RP training was introduced to bring this decisive weapon into the RAAF's operational armoury; the Beaufighter with its long range, low-level speed and approach was the ideal weapon-carrier. Strikes would now begin against enemy installations, in the Dutch East Indies, the elimination of Japanese garrisons still holding out, and the strafing of all enemy seaborne traffic.

The 10 Operational Group was reformed into the 1st TAF at Noemfoor on 25 October 1944, under the command of Air Commodore A. Colby, and included both the Beaufighter squadrons amongst its units. Its main role now was the destruction of the enemy transports and supplies, and with the coming invasion of the Philippines, to fully occupy the Japanese in the Dutch East Indies; the Americans having decided that *only* the Americans would make a triumphant return to the Philippines.

During the first week in November, bad weather hampered operations by 1st TAF, but by the second week advance units of 77 Wing moved to Morotai. 30 Squadron moved to Morotai on 16 November, and on the 22nd strafed coastal traffic in the Davao Gulf; the squadron having lost A19-206 at Kamiri during take-off and A19-186 during landing at Wama. The night of the 22nd found the Japanese retaliating, attacking Pitu and Wama, destroying amongst other aircraft, five Beaufighters — A19-179, 190, 196, 199 and 214. Two nights later in another raid on Pitu, Beaufighters A19-29 and -35 were destroyed.

During these raids a number of Bostons were also destroyed, and due to the shortage of this type of aircraft 22 Squadron was withdrawn to Noemfoor to re-equip with Beaufighters, the squadron commanded by Wing Commander C.E. Woodman. The Squadron would not however commence operations with Beaufighter until 1945.

140

31 Squadron and its Beaufighters led by Squadron Leader J. Boyd, moved to Morotai in December, where it joined 30 Squadron. The squadron's first operation was flown on 9 December, when fourteen of their aircraft attacked Jolo Island in the Philippines. Going in fast and low, the Beaufighters strafed troops and dockside buildings and were well plastered with ack-ack in return, with A8-7 landing at base on return and swinging off the runway — the RAAF were to suffer an amount of this with their Australian-built Beaus, which had no tailwheel lock fitted.

The next three days saw the two Beaufighter squadrons combined to carry out attacks on targets in the Celebes. Then on the 22nd–25th of the month they subjected installations on Halmahera to a four-day blitz, flown in conjunction with Kittyhawks of 80 Squadron — the Kittyhawk now being capable of carrying a 1500-lb bomb-load. Over the four days there were 384 Kittyhawk sorties flown and 129 Beaufighter ones.

January 1945 found the three Beaufighter squadrons, 22, 30 and 31 in the 77 Wing of 1st TAF RAAF and under the operational control of the 13th US Air Force, attacking bypassed enemy garrisons or targets. This direction by the 13th US Air Force to 1st TAF to clean up bypassed Japanese targets instead of partaking in the main attack against the Japanese, resulted in an amount of anger, as it was seen as the USA Services grabbing the glory. In the end a number of senior RAAF officers objected to this policy in forceful terms, and 1st TAF was given the chance to take part in the assault on Tarakan, Borneo, Labuan and Balikpapan.

In the meantime, on 5 January 1945, twenty-one Beaufighters of 30 and 31 Squadrons dropped napalm as well as high explosive bombs in the Menado area, this being the first use of napalm by RAAF squadrons. The role of the Beaufighters was now the neutralization and destruction of the Japanese forces in the Celebes, Ceram and Ambon areas. On 1 February striking at Tomohon troop-holding area, eight Beaus strafed personnel, but met with a reception of fierce concentrated light ack-ack, which claimed A8-10 and A8-32 over the target.

22 Squadron flew their first Beaufighter operation on 11 February, when twenty-eight aircraft from 22, 30 and 31 Squadrons, along with Kittyhawks, attacked Tondano in the Celebes. 22 Squadron were to lose their first Beau on operations, when four of their aircraft attacked a bridge at Isumu, near Gorontal. A8-43 flown by Squadron Leader J. Holloway and

Mk 21 named 'Bridgetown' A8-247 on strip on Los Negros, Admiralty Islands, 1945. (Frank Smith)

Flying Officer D. Genders was seen to hit a tree near the bridge and crash; Holloway was killed and Genders died as a PoW.

Another Beaufighter unit came into existence on 22 January 1945, when 93 Squadron was officially formed at Kingaroy under Wing Commander D. Gulliver. The squadron joined 1st TAF at Morotai at the end of April; by now Morotai was getting a little crowded and 93 Squadron had moved to Labuan by 13 June.

The 1st Australian Corps were ordered to occupy Tarakan by 1 May, so air strikes were made against enemy bases, and on 20 April four Beaufighters of 31 Squadron escorted a USAAF Mitchell on a low-level photo-recce along the Tarakan waterfront. This met with little opposition, and the actual landing on Tarakan was made against light opposition and few casualties. The Beaufighters of 77 Wing (22, 30 and 31 Squadrons) moved to Sanga-Sanga airfield on 15 May to give night-fighter cover over the Tarakan area.

Labuan dominated the entrance to Brunie Bay and was the next invasion area, so that at the start of May the initial strikes began, the Beaus of 77 Wing to the fore making attacks on targets in Labuan and the waters around it. The Japanese Army appeared to have given up the area in general and 'trade' was in short supply. Despite the torrential rain and the rest of life's hazards in a war zone, the Airfield Construction Unit rebuilt the airfields at Tarakan, whilst 77 Wing (22, 30 and 31 Squadrons) operated from Sanga-Sanga to give fighter cover over the area.

Early in May the air assaults preceding the invasion of Labuan began, with 30 Squadron on 9 June losing A8-177 during a low-level attack on Beaufort, the aircraft hitting a tree during the strike and ploughing in. The following day Australian troops landed at Labuan and quickly secured the airstrip for fighter operations. On the 11th 31 Squadron were to lose two Beaufighters in a mid-air collision off Sanga-Sanga, when A8-39 and A8-191 collided with each other and crashed into the sea. On the same day, 22 Squadron had A8-42 crash on take-off — this was not due to any peculiarity of the Beaufighter, for other squadrons with other types of aircraft were also suffering such losses; the main reason was the primitive nature of the airstrips and airfield control.

Meanwhile 93 Squadron with its Beaufighters had come under the control of 86 Wing, which had moved from Morotai to Labuan. With the war virtually over, 93 Squadron only took part in two operations, losing A8-85 on 7 August during an attack on shipping at the mouth of the Tabuan River. The other Beau squadrons flying their last 'in anger' operations on 1 August and thereafter consoled themselves with recce sorties from Morotai and Tanitini, the airfield at Tarakan proving far from ideal for the Beaufighter.

The Mosquito aircraft were by now intruding into the operational scene, but the Beaufighter squadrons of the RAAF were to stay strictly Beau until the cessation of the Pacific War on 15 August 1945. This was fortunate for them, for

Mk IVc A19-171 of 30 Squadron on Tadje strip, New Guinea, August 1944. (Frank Smith)

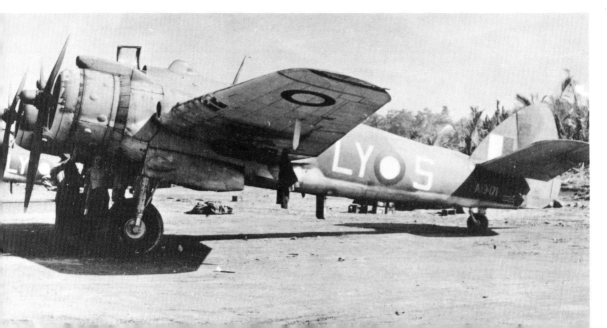

Good underneath view of A8-358 as it banks away, showing torpedo crutches fitted. (Frank Smith)

Beaufighter A8-99 flying over Melbourne; note autopilot bulge in front of windshield. (Keith Meggs)

Mk TT.10 at Seletar, Far East, 1956. (Paddy Porter)

wooden structures in the hot humid climate and primitive conditions of the SWPA were far from ideal. The Beau as an 'interim' design, had given to the men of the RAAF a strong, long-range strike aircraft, which they wielded with power against the Japanese — a foe who fought fiercely but without humanity, and because of this many RAAF personnel died after capture.

In similar circumstances to the RAF, the RAAF after hostilities had ceased, relegated the Beaufighters to the scrap heap. Yet in no other country are there more Beaufighters restored or being restored, than in Australia. The Beau after all was typical of Australia, tough and a hard hitter, and she gave to the RAAF the power to strike back.

Postcript

The Beaufighter was to continue into the post-war years, both in an operational and training capacity. In the training role the Beaufighter was used as a target-towing aircraft; first of all in Australia in late 1945 where A8-265 was converted with the installation of a target-towing winch and its related equipment — two complete years before the RAF carried out the first airtest of NT913, the prototype TT.10.

In November 1947 approval was given for the TT.10, thirty-five airframes being converted from Mk. Xs at Filton. The first conversion was NT913, with its first flight in May 1948. The TT.10 differed from its stable-mates in having no armour, armament or blast tubes fitted, and the rear lower escape hatch was no longer usable, and was only suitable for the launching of the towed target. The observer was replaced by a winch operator, and there was no provision for a third crew member. The aircraft had the same tankage as other marks, plus fuel tanks in the wing gun positions, but with the self-sealing covering of the tanks removed and replaced by packing, the fuel tank capacity was 682 gallons.

The TT.10 was powered by two Hercules 18 engines, which were fitted with a two-speed supercharger. Around the tailplane was fitted a cable-guard. These aircraft were delivered to the RAF over a period of two years; commencing in 1948, they were flown by co-operation units on gunnery practice over a number of areas, notably Gibraltar, Cyprus, Malta and Malaya. The last one in service was RD761, which was flown at Seletar, Singapore, and made its last flight there on 16 May 1960 — nearly twenty-one years after the first Beaufighter flew — as an interim design!

In Australia the Beaufighter continued in limited use with the RAAF, with the last reported flight on 9 December 1957, when A8-357 of the No. 1 Air Trials Unit was ferried for scrapping. Thanks to a number of enthusiasts in Australia there are some Beaufighters restored or in a state of restoration. The Camden Museum of Aviation has completed the restoration of Australian-built A8-186, and is restoring the cockpit section of A8-386; whilst the Australian Aircraft Restoration Group at Moorabbin has an almost completed A8-328, the rear fuselage of A8-384 and the fuselage of A19-43. A further search by enthusiasts in 1981 discovered a further two, A19-144 and A19-148, on a disused wartime air strip at Drysdale River Mission; these were transported by road to Sydney, where a group of enthusiasts are renovating one.

On VE-Day (8 May 1945), there were still fifteen Beaufighter Squadrons scattered around in the various battle areas covered by the Royal Air Force, as well as numerous units that had a few Beaufighters at their establishment, plus numerous instructional airframes around the training schools. The RAAF were still operating four squadrons in the South-West Pacific Area. By VJ Day on 15 August however, quite a number of RAF squadrons were ready for disbanding or for conversion to Mosquitoes.

One unit, 27 Squadron, which had been designated an Air Jungle Search and Rescue unit, was, by September, under the threat of disbandment. However, with the civil strife in the Dutch East Indies from insurgents trying to form their own idea of independent states, this disbandment was put into reverse. The result was that the squadron was employed over a wide area on a number of tasks, and was eventually disbanded on 1 February 1946.

89 and 176 Squadrons continued in the Far East as the surrender took effect in remote areas, and as civil order was once more established, exchanging their Beaufighters for Mosquitoes, and in the case of 89 Squadron, being renumbered to 22 Squadron. So that by the middle of 1946 there were no Beaufighter squadrons operating in the Far East; however, there came an about-face, not quite so surprising considering that the Beaufighter had been so satisfactory, and by the end of the year the tough old Beau was back, with 84 Squadron exchanging their Mosquito FBVIs for Beaufighter TF.Xs; 45 Squadron did the same shortly afterwards. In the meantime 22, 176, 177 and 211 Squadrons had been disbanded.

Mk TF.X of 42 Squadron at London airport, 1946.

45 and 84 Squadrons continued operating their Beaufighters in a number of roles, with both squadrons operating against Chinese Communist terrorists in Malaya during the years up to 1949 in 'Operation Firedog'. Just afterwards they started converting to the Bristol Brigand, a design which continued the role of the Beaufighter.

Back in the UK, the last two Coastal Command squadrons, 144 and 455, were disbanded in May 1945, and it appeared as if this was the end of the Beaufighter as a ship-strike weapon. Once again a Beaufighter squadron was reborn, when on 1 October 1946 at Thorney Island, 42 Squadron was re-formed as a torpedo-bomber unit equipped with TF. Mk. X aircraft. This time its life was short, as the unit was disbanded again, approximately twelve months later.

A number of countries had been, and were, interested in purchasing the Beaufighter; one of the most obvious was Turkey, which had already acquired a number locally during the war. Then in 1946 the country placed an order for twenty-four refurbished aircraft from Bristol, all ex-RAF machines that were delivered in early 1947.

A8-186 restored by the Camden Museum of Aviation, Australia. (G. Dinsdale)

Portugal was another purchaser, placing an order for sixteen of the TF. Mk. X, again all were refurbished ex-RAF aircraft, and were to replace the Blenheim IV aircraft supplied during the war to Portugal for that country's defence. It was one of these Portuguese aircraft, ex-RD253, which the Portuguese Government presented back to the RAF.

The Dominican Republic also purchased ten refurbished Beaufighter VI aircraft, which were converted back from TF.X machines. Israel was another customer, not only unofficially but in an indirect way; for with an embargo placed on the selling of warlike material to the Middle East, Israel was forced to use any method to acquire means of defence against her Arab neighbours. In the case of the Beaufighters, six had been acquired and refurbished by Fairey Aviation, and converted with civil registrations; whether there would have been a civil use for them is open to opinion — though one could say the same for civilian Mosquitoes. Five were legally purchased for use by a supposed film-making company, four disappeared whilst *en route* to a 'filming' location in Scotland and the fifth crashed at Thame. The four, as is well documented, arrived in Israel and were used in Israel's fight for survival. Due to lack of spares and engines, and shortage of maintenance, the Beaus proved hard

to keep serviceable and appear to have disappeared in the maelstrom of war at the time.

One of the units operating the TT.10 was based at Malta, and after the unit had closed down one TT.10 could be seen at Luqa Airport, left gradually deteriorating and derelict. Many spoke their thoughts about resurrection of the old wreck, but it was not until 1963 that it was transported to Britain for restoration. At approximately the same time a Beaufighter was required by the RCAF for the National Museum of Canada; now there was the possibility of two Beaufighters in the RAF Museum, RD253 ex-Portugal and RD867 ex-Malta, so an agreement was made to 'swop' RD867 for a Canadian-made Bolingbroke (Blenheim IV). This entailed the restoration of RD867 and transportation to Canada; whilst down at St Athan RAF enthusiasts would commence the restoration of RD253 once it was transferred there. This was done and a thorough renovation of airframe and engines has resulted in a complete and pristine-condition Beaufighter being on public display at the RAF Museum, Hendon.

Although a small number of Beaufighters were acquired by France after 1945, little is known of their use, apart from the carriage of missiles for test and experimental work. Bristol's also maintained one for the development of their Hercules engines and power plants. At Farnborough the RAE had a Mk. II T3032, which was in use at the SME Flight (Structures and Mechanical Engineering) until August 1945, when its airframe hours were finished and so it was dismantled — it had made its first appearance at a Ministry (A&AEE) establishment in January 1943.

Post-war British ordnance developed a copy of the German Mauser 40 mm aircraft cannon, which they named the Aden gun. The initial air firing trials of the 30 mm Aden gun was carried out on Beaufighter RD388 in August 1951. The gun had a shortened barrel and was mounted in the port gun bay, mounted on its side with the right-hand side of the gun face downwards, and had a left-hand feed. The flight trials took place on the Lyme Bay range, with shallow dives made from 1200 ft and the aircraft speed at the point of firing being 275 knots (320 mph). The first clear shoot of 100 rounds in three bursts took place on 29 August. A further series of firing trials were undertaken in December 1951 and in January 1952, when the Beaufighter mounted two Aden cannon. As on the initial trials the ejected cartridge links had damaged the tailplane and elevator; cartridge-link collectors were fitted for the remainder of the trials.

As with most aircraft books the text has centred around the aircraft, flying and aircrew, yet it must not be forgotten that the success of every flight or combat depends on the backing of the groundcrew, that without the groundcrew there would be no flying; without the 'erk' nothing would get done on the station or squadron. Whether in the Far and Middle East, or Europe, these were the men who kept the aircraft going and the aircrew flying; nearly always working in the most atrocious conditions in mud, rain or sand, and always the worst accommodation; quite often working under wartime engineering officers who, though they might have qualified at a University, were as thick as two short planks when it came to basic engineering and man-care. The 'erk' was the last in line when the buck had to be passed, expected to work under any conditions — yet be presentable when the AOC decided to pop in. These were the men that were the backbone of any squadron, the ones who kept the Beaufighters flying — and the Beaufighter will live on in the memories of the groundcrew, as well as the aircrew. She was one of the most outstanding aircraft of World War II, her crews had complete confidence in her durability, and the strength of her construction would prove the salvation of many of them; praise and affection go hand in hand when one talks of the Beaufighter — she was tough, she was Bristol-designed, she was the MIGHTY BEAU.

Abbreviations

A&AEE	Aeroplane and Armament Experimental Establishment		MAP	Ministry of Aircraft Production (or Minister)
AD/RDL	Assistant Director/Research and Development (section L)		MU	Maintenance Unit
			NACA	National Advisory Committee on Aeronautics (USA)
AFDU	Air Fighting Development Unit			
AGME	Aircraft Gun Mounting Establishment		NJG	*Nachtjagdgeschwader* (night fighter unit)
ASV	Air to surface vessel (radar)		PV	Private venture
ATDU	Aircraft Torpedo Development Unit		RAAF	Royal Australian Air Force
CG	Centre of gravity		RAE	Royal Aircraft Establishment (now Royal Aerospace Establishment)
CO	Commanding Officer			
DAD	Director of aircraft development			
DD/RD	Deputy Director/research & development		RCAF	Royal Canadian Air Force
			R&M	Research and Memorandum
			RNZAF	Royal New Zealand Air Force
DFC	Distinguished Flying Cross		RO	Radar operator
DFM	Distinguished Flying Medal		RP	Rocket projectile
DGRD	Director General Research and Development		RTO	Resident technical officer (of MAP)
			SWG	Standard wire gauge (metal thickness measurement)
DSO	Distinguished Service Order			
DTD	Director of Technical Development		TAF	Tactical Air Force
EA	Enemy aircraft		TFU	Telecommunications Flying Unit
FIU	Fighter Interception Unit		TRE	Telecommunications Research Establishment
GCI	Ground controlled interception			
KG (KGr)	*Kampfgeschwader* or *Gr Gruppen* (bomber unit)		TT	Target Towing
			VCAS	Vice Chief of Air Staff
LAC	Leading aircraftsman		VGO	Vickers gas operated (gun) Type 'K'

Appendix 1
British Early Warning Radar

The potential threat of air attack was taken so seriously in the mid-1930s, that a system was designed, developed and was in limited operational use by 1938. Known under the cover name of RDF (radio direction finding), it was not until 1943 that the American term 'Radar' (derived from **R**adio **A**id to **D**irection **a**nd **R**anging) was adopted, but it must be understood that the two terms are synonymous.

The Chain Home (CH) station's sole function was early warning, using horizontally-polarized aerials which radiated pulsed transmissions on a wavelength of 10 metres. The continuous train of pulses radiated in an arc of 180 degrees, to form a broad 'floodlight' extending in front of each station towards the continent. As a protection against interference or deliberate jamming, provision was made for switching to pre-selected frequencies. Presentation was made in the use of a CRT receiver display as opposed to the PPI (Plan Position Indicator) trace; and stations could give bearing and distance with calculated altitude, but CH was not reliable for targets below 10,000 feet. With a power output of 200 kW the maximum range was in excess of 120 miles.

With the need for lower-altitude cover all too obvious, work at Bawdsey adapted a system of ship location to provide a back-up to CH; this system was the generation and reception of transmissions on 1.5 metres. It had a steerable aerial array and used a 'split-beam' technique which could give an accurate bearing to within a few minutes of arc, and this down to 500 feet. This system was known as Chain Home Low (CHL) and came into operation in 1940.

With mounting concern about mine-laying aircraft, low-flying raiders, E-boats, and even U-boats, a further system, known as Chain Home Extra Low (CHEL), was introduced; thus providing a three-layer system by which the RAF controlled all fighter interception through Ground Controlled Interception (GCI). The CHEL system used the Army gun-control radar that the RAF took over, and which operated on 10-centimetre wavelength. This radar was based on the cavity magneton (a highly secret valve, which could generate very high frequency output at very high power), and presentation was on PPI screens.

The information from the CH stations was however not in a form suitable for Fighter Command, which used a Grid Plotting system, so an early analogue computer was constructed, which enabled operators to key-in information at their desks and RAF Sector Control then received grid position, height and estimated strength of a raid from the Filter Room.

With the rapid expansion of RDF work it was necessary to diverge the operation into two separate, yet interlinked, branches; these were RDF1 for ground-based systems and RDF2 for airborne systems. RDF1 came into operational use in 1938, with the first five CH stations covering the approaches to the Thames Estuary, and were quite prominent around the eastern and south-eastern coast of Britain, with 350-ft high steel lattice towers, usually sited in groups of three. By September 1939 the cover had been extended from the Isle of Wight to the Orkneys, with only one gap, which was around Aberdeen. Thus the RAF became the first to incorporate such a system into its defensive structure, and the chain was able to distinguish between friend and enemy by the use of IFF sets fitted in the RAF aircraft.

RDF2 gave to Great Britain the privilege of being the first nation to embark on a system of airborne radar prior to the 1939–45 war, though this was at the time quite crude and the sets 'handmade'. Airborne interception radar in Great Britain quickly developed through from the Mk. I to Mk. IV, the latter being the first operational set in the true sense of the word. From there it developed to the Mk. VII, then with the introduction of the magnetron valve and 10-cm wavelength came the Mk. VIII and IX. In the meantime the secrets of the magnetron valve had been communicated to the USA, where work began on AI sets using the magnetron valve at MIT, which culminated in the SCR520; ten sets of this type were ordered by the British Government in September 1941 for installation in Beaufighters. This was later cancelled when it was found that the equipment would not fit into the Beaufighter's nose. From there hybridization

took place between the AI Mk. VIII and SCR520, that resulted in the SCR720, known in the RAF as AI Mk. X.

With the co-operation of the British the USAAF were able to create a specialist night-fighter force, which was well advanced by December 1941, but it would not be until September 1942 that a radar-equipped night-fighter in the USAAF reached a front-line squadron, and that was not in Europe. The European theatre of operations found four USAAF squadrons equipping with Beaufighter VIF aircraft with AI Mk. IV.

Appendix 2
Aircraft Serial Numbers

1 Produced by the Bristol Aeroplane Co Filton

Type	Quantity	Manufacturer's serial numbers	RAF serial numbers
proto	6	9562 – 9567	R2052 – 2057.
Mk. 1	2	9569 – 9570	R2059 – 2060.
Mk. 1	139	9573 – 9711	R2063 – 2101, 2120 – 2209, 2180 – 2209, R2240 – 2269.
Mk. 1	78	10194 – 10271	T3228 – 3250, 3270 – 3272, 3290 – 3333, T3348 – 3355.
Mk. 1	125	10419 – 10543	V8219 – 8233, 8246 – 8289, 8307 – 8356, V8370 – 8385.
Mk. II	1	9568	R2058.
Mk. II	2	9571 – 9572	R2061 – 2062.
Mk. II	150	9712 – 9861	R2270 – 2284, 2300 – 2349, 2370 – 2404, R2340 – 2479.
Mk. II	150	10044 – 10193	T3009 – 3055, 3070 – 3107, 3137 – 3183, T3210 – 3227.
Mk. II	147	10272 – 10418	T3356 – 3389, 3410 – 3447. V8131 – 8170, 8184 – 8218.
Mk. VIF	418	10544 – 10961	V8386 – 8419, 8433 – 8472, 8489 – 8528, V8545 – 8594, 8608 – 8657, 8671 – 8720, V8733 – 8778, 8799 – 8848, 8862 – 8901, BT286 – 303.
Mk. VIF	100	11684 – 11783	MM838 – 887, MM899 – 948.
Mk. VIF	150	12305 – 12454	ND139 – 186, 198 – 243, 255 – 299, ND312 – 322.

2 Produced by Fairey Aviation, Stockport *

Type	Quantity	RAF serial numbers
Mk. IF	25	T4623 – 4647.
Mk. IC	300	T4648 – 4670, 4700 – 4734, 4751 – 4800, 4823 – 4846, T4862 – 4899, 4915 – 4947, 4970 – 5007, 5027 – 5055, T5070 – 5099.
Mk. VIC	175	T5100 – 5014, 5130 – 5175, 5195 – 5200, 5250 – 5290, T5315 – 5352.

3 Produced by Rootes, Blythe Bridge

Type	Quantity	RAF serial numbers
Mk. VIF	150	KV896 – 944, 960 – 981, KW101 – 133, 147 – 171, 183 – 203.
Mk. X	110	KW277 – 298, 315 – 355, 370 – 416.

4 Bristol/MAP Shadow Factory, Old Mixton, Weston

Type	Quantity	RAF serial numbers
Mk. IF	240	X7540 – 7589, 7610 – 7649, 7670 – 7719, 7740 – 7779, X7800 – 7849, 7870 – 7879.
Mk. VIF	260	X7880 – 7899, 7920 – 7924, 7926 – 7936, 7940 – 7969, X8000 – 8029, 8100 –8109, 8130 – 8169, 8190 – 8229, X8250 – 8269. EL145 – 192, 213 –218.
Mk. VIC	518	X7925. X7937 – 7939, 8030 – 8039, 8060 – 8099. EL219 – 246, 259 – 305, 321 – 370, 385 – 418. EL431 – 479, 497 – 534. JL421 – 454, 502 – 549, 565 – 582, 584 – 592, 619 – 628, JL639 – 648, 659, 704 – 712, 723 – 735, 756 – 779, JL812 – 826, 836 – 855, 869 – 875.
Mk. VI (ITF)	60	JL583. 593. 610 – 618, 629 – 638, 649 – 658, 713 – 722, JL827 – 835, 949 – 957. JM104.
Mk. XI	163	JL876 – 915, 937 – 948. JM105 – 136, 158 – 185, 206 – 250, 262 – 267.
Mk. X	2095	JM268 – 291, 315 – 356, 379 – 417. LX779 – 827, 845 – 887, 898 – 914, 926 – 959, 972 – 999. LZ113 – 158, 172 – 201, 215 – 247, 260 – 297, 314 – 346, LZ359 – 384, 397 – 419, 432 – 465, 479 – 495, 515 – 544.

* See RAAF serial numbers and transfers.

NE193 – 232, 245 – 260, 282 – 326, 339 – 386, 398 – 446,
NE459 – 502, 515 – 559, 572 – 615, 627 – 669, 682 – 724,
NE738 – 779, 792 – 832.
NT888 – 929, 942 – 971, 983 – 999.
NV113 – 158, 171 – 218, 233 – 276, 289 – 333, 347 – 390,
NV413 – 457, 470 – 513, 526 – 572, 585 – 632.
RD130 – 176, 189 – 225, 239 – 285, 298 – 335, 348 – 396,
RD420 – 468, 483 – 525, 538 – 580, 685 – 728, 742 – 789,
RD801 – 836, 849 – 867.
* SR910 – 919.

5 Australian Department of Aircraft Production

Type	Quantity	RAF serial numbers
Mk. F21	364	A8–1 to A8–364.

6 Fairey-produced aircraft transferred to RAAF

T4920 – 4931	became A19–1 to A19–12.
T4943 – 4947	became A19–13 to A19–17.
T4970 – 4978	became A19–18 to A19–26.
T4991 – 5004	became A19–27 to A19–40.
T5047 – 5055	became A19–41 to A19–49.
T5070 – 5074	became A19–50 to A19–54.
T5075	became A19–63
T5076	became A19–55
T5077	became A19–61
T5081	became A19–56
T5082	became A19–64
T5083 – 5084	became A19–57 to A19–58
T5086	became A19–71
T5089	became A19–62
T5090 – 5091	became A19–59 – A19–60
T5092	became A19–72
T5093	became A19–66
T5094	became A19–65
T5095	became A19–67
T5097	became A19–68
T5098	became A19–70
T5099	became A19–69
T5200 –5201	became A19–73 – A19–74
T5202	became A19–77
T5203	became A19–76
T5204	became A19–78
T5205	became A19–75
T5254	became A19–86
T5255	became A19–88
T5257	became A19–89
T5262	became A19–85

* Note: SR911 – 914, 916 – 917, 919 converted to TT.10 aircraft.

T5263	became A19–97
T5264	became A19–87
T5270	became A19–90
T5295 – 5296	became A19–98 – A19–99
T5327	became A19–111
T5328	became A19–110
T5329	became A19–115
T5330	became A19–112
T5331	became A19–114
T5336	became A19–119
T5337	became A19–124
T5338	became A19–120
T5339	became A19–126
T5340	became A19–128
T5341	became A19–127
T5342	became A19–129
T5343 – 5344	became A19–135 – A19–136

7 Old Mixon-produced aircraft transferred to RAAF

EL241	became A19–94
EL243	became A19–79
EL244	became A19–92
EL245 – 246	became A19–80 – A19–81
EL259	became A19–84
EL260 – 261	became A19–82 – A19–83
EL395 – 396	became A19–101 – A19–102
EL411	became A19–107
EL412	became A19–105
EL413	became A19–100
EL435	became A19–96
EL436	became A19–91
EL438	became A19–93
EL443	became A19–95
EL510	became A19–106
EL518	became A19–109
EL520	became A19–108
JL429	became A19–104
JL430	became A19–103
JL582	became A19–116
JL584	became A19–113
JL826	became A19–118
JL837	became A19–123
JL838	became A19–122
JL839	became A19–121
JL840	became A19–125
JL841	became A19–117
JL847	became A19–132
JL848	became A19–130
JL850	became A19–133
JL853	became A19–131
JL854	became A19–134
JL946	became A19–148

JM120	became A19–152
JM131	became A19–158
JM134 – 135	became A19–143 – A19–144
JM164	became A19–163
JM165	became A19–161
JM167	became A19–160
JM168	became A19–141
JM169	became A19–146
JM170	became A19–156
JM171	became A19–147
JM175	became A19–137
JM176	became A19–151
JM177	became A19–142
JM182 – 184	became A19–138 – A19–140
JM185	became A19–145
JM236	became A19–157
JM272	became A19–159
JM273	became A19–162
JM282 – 284	became A19–153 – A19–155
JM285 – 286	became A19–149 – A19–150
LX988 – 995	became A19–164 – A19–171
LZ195 – 201	became A19–172 – A19–178
LZ215	became A19–179
LZ321 – 328	became A19–180 – A19–187
NE229	became A19–189
NE230	became A19–197
NE231 – 232	became A19–191 – A19–192
NE245	became A19–193
NE356	became A19–198
NE357	became A19–188
NE358 – 359	became A19–194 – A19–195
NE360	became A19–190
NE361	became A19–196
NE482	became A19–199
NE483	became A19–202
NE484	became A19–201
NE485	became A19–200
NE487	became A19–203
NE573 – 574	became A19–206 – A19–207
NE584 – 585	became A19–204 – A19–205
NE586 – 589	became A19–208 – A19–211
NE693 – 696	became A19–212 – A19–215

Appendix 3
Airframe Data

	Mk. II	Mk. VI	Mk. X
Wingspan	57 ft 10 in	57 ft 10 in	57 ft 10 in
Wing area, total	451 sq ft	451 sq ft	451 sq ft
Aerofoil section	RAF 28 basic	RAF 28 basic	RAF 28 basic
Incidence	$2\frac{1}{2}$ degrees	$2\frac{1}{2}$ degrees	$2\frac{1}{2}$ degrees
Dihedral	6 degrees 30 min	6 degrees 30 min	6 degrees 30 min
Tailplane span (early) (later)	18 ft 4 in 20 ft 5 in	20 ft 5 in	20 ft 5 in
Tailplane area (early) (later)	86.4 sq ft 101.75 sq ft	101.75 sq ft	101.75 sq ft
Incidence	nil	nil	nil
Dihedral (early) (later)	nil (early) 12 degrees (later)	nil 12 degrees	12 degrees
Elevator area (early) (later)	32.6 sq ft 38.5 sq ft	38.5 sq ft	38.5 sq ft
Fin area	13 sq ft	13 sq ft	13 sq ft
Rudder area	25.7 sq ft	27 sq ft	27 sq ft
Flap area, total	57.1 sq ft	57.1 sq ft	57.1 sq ft
Aircraft length	42 ft 6 in	41 ft	41 ft 4 in
Undercarriage track	18 ft	18 ft	18 ft
Undercarriage legs	Lockheed oleo pneumatic		
Tailwheel unit	Lockheed oleo pneumatic	oleo pneumatic or BLG	BLG oleo pneumatic

Engines	R-R Merlin XX	Hercules VI or XVI	Hercules XVII
Propellers	Rotol VP RS5/5 12 ft dia	DH Bracket 6/6 12 ft 9 in dia	DH Hydromatic 55/3 or 55/14 or 55/12 or Bracket 6/6 12 ft 9 in dia
Fuel capacity, normal	550 gallons	550 gallons	682 gallons
overload	624 gallons	682 gallons	882 gallons
Weight, tare	13,800 lb	14,875 lb	15,592 lb
loaded	21,000 lb	22,779 lb	25,400 lb
Maximum speed and altitude	337 at (mph and ft) 21,000	337 at 16,000	320 at 10,000
Maximum speed at SL		295 mph	305 mph
Service ceiling		26,000 ft	19,000 ft *
Range, normal		1540 mile	1400 mile

* Engines rated for low and medium altitudes.

Appendix 4
Engine Data

1 Bristol Hercules

	Hercules VI	Hercules XVII
Bore	5.75 in	5.75 in
Stroke	6.5 in	6.5 in
Cubic capacity (swept volume)	2366 cu in	2366 cu in
Compression ratio	7.0 to 1	7.0 to 1
Supercharger gear ratio MS	6.679 to 1	6.679 to 1
FS	8.365 to 1	8.365 to 1 *
Supercharger impeller dia	13.25 in	12.0 in
Propeller shaft size	SBAC No. 6	SBAC No. 6
Reduction gear ratio	0.444 to 1	0.444 to 1
Carburettor type	Hobson AIT 132MC	Hobson AIT 132ME or 132 MF
Fuel specification octane	100/130	100/130
Oil specification	DED 2472	DED 2472
Weight, nett dry	1930 lb	1930 lb
Maximum take-off rpm & boost	2800 +8$\frac{1}{4}$ psi	2900 +8$\frac{1}{4}$ psi
Maximum take-off power	1615 bhp	1725 bhp
Maximum climb rpm & boost	2400 +6 psi	2400 +6 psi
Maximum climb power MS	1355 bhp at 4750 ft	1395 bhp at 1500 ft
FS	1240 bhp at 12,000 ft	....
Maximum continuous cruise rpm and boost	2400 +6 psi	2400 +6 psi

* FS gear inoperative on Hercules XVII.

Maximum continuous cruise MS	1355 bhp at 4750 ft	1395 bhp at 1500 ft
FS	1240 bhp at 12,000 ft	....
Maximum ECB rpm and boost	2400 +2 psi	2400 +2 psi
Maximum economical cruise MS	1050 bhp at 10,250 ft	1085 bhp at 7000 ft
FS	955 bhp at 17250 ft	....
Maximum oil temperature on takeoff	90°C	90°C
Maximum cylinder head temperature at start of take–off	230°C	230°C
Maximum oil temperature on climb	90°C	90°C
Maximum cylinder head temperature on climb	270°C	270°C

2 Rolls-Royce Merlin XX

Bore	5.4 in
Stroke	6.0 in
Cubic capacity (swept volume)	1649 cu in
Compression ratio	6.0 to 1
Supercharger gear MS	8.15 to 1
FS	9.49 to 1
Supercharger impeller dia	10.25 in
Propeller shaft size	SBAC No. 5
Reduction gear ratio	0.42 to 1
Carburettor type	SU AVT40/214 or 216
Fuel specification octane	100/130
Oil specification	DED 2472
Weight, nett dry	1450 lb
Maximum take-off rpm and boost	3000 +12 psi
Maximum take-off power	1280 bhp
Maximum climb rpm and boost	2850 +9 psi
Maximum climbing power MS	1220 bhp at 9750 ft
FS	1130 bhp at 16,500 ft

Maximum continuous rpm and boost	2850 +9 psi
Maximum continuous power MS	1220 bhp at 9750 ft
FS	1130 bhp at 16,500 ft
Maximum ECB rpm and boost	2650 +7 psi
Maximum economical cruise MS	1080 bhp at 8750 ft
FS	1015 bhp at 15,500 ft
Maximum oil temperature at take-off	105°C
Maximum coolant temperature at take-off	135°C
Maximum oil temperature in cruise	90°C
Maximum coolant temperature in cruise	105°C

Appendix 5
Beaufighter Squadrons

Operational squadrons that flew Beaufighters, and their code letters

RAF
22
23 ZK
27
29 RO
39 XZ
42 QM
45 OB
46 FH
47 KU
68 WM
84 PY
89
96 ZJ
108 LD
125 VA
141 TW
143 HO
144 PL
153 TB
176
177
211 UQ
217 MW
219 FK
227
235 LA
236 MB
248 WR
252 PN
254 QM
255 YD
256 JT
272 L7
287 KZ
307 EW
515 3P
577 3Y
600 BQ
603 XT
604 NG

RAAF	455	UB
RAAF	456	RX
RCAF	404	EE
RCAF	406	HU
RCAF	409	KP
RCAF	410	RA
RNZAF	488	ME
RNZAF	489	P6
RAAF/SWPA	22	DU
RAAF/SWPA	30	LY
RAAF/SWPA	31	EH
RAAF/SWPA	92	OB
RAAF/SWPA	93	SK
SAAF	16	
SAAF	19	
USAAF	414	
USAAF	415	
USAAF	416	
USAAF	417	

Although these code letters were allotted to the squadrons, the aircraft did not always carry them.

Bibliography

Barnes, C.H. *Bristol Aircraft Since 1910* (Putnam)

Beaufighter Air Publication 1721 (HMSO)

Beaufighter Test Reports (A&AEE/758)

Bergel, Hugh, *Fly and Deliver* (Airlife Publishing Ltd)

Bowen, E.G. *Radar Days* (Adam Hilger)

Bowyer, Chaz *Beaufighter* (William Kimber)
    Beaufighter at War (Ian Allan)
    The Flying Elephants (Macdonald)

Braham, J. *Scramble* (W. Kimber)

Chisholm, R. *Cover of Darkness* (Chatto & Windus)

Coastal Command (HMSO)

Cunningham, A. *Tumult in the Clouds* (P. Davies)

Gunston, Bill *Night Fighters* (Patrick Stephens)

Halley, J. *RAF Unit Histories* (Air Britain)

Hay, M.M.D. and Scott, J.D. *History of World War II — Design and Development of Weapons* (HMSO)

Hornby, W. *History of the Second World War* (HMSO)

Innes, D. *Beaufighters Over Burma* (Blandford)

Nesbit, R. *The Strike Wings* (W. Kimber)

RAAF Official History (AWM)

RAF Middle East (HMSO)

Rawlings, J. *Coastal and Support Squadrons of the RAF* (Janes)

Rawnsley & Wright *Night Fighter* (Collins)

Robinson, A. *Night Fighter* (Ian Allan)

Shores, Ring & Hess *Fighters Over Tunisia* (N. Spearman)

Sword, S.S. *Technical History of the Beginnings of Radar* (Peter Peregrinus Ltd)

Taylor, H. *Test Pilot at War* (Ian Allan)

Wallace, G.F. *Guns of the RAF* (W. Kimber)

The above is suggested reading with regard to the Beaufighter, its operations, territorial scenes and background. Detailed reading can be carried out in conjunction with records at the Public Record Office, Kew, under catalogue headings AIR2, AIR19, AIR20, AIR24–27, AIR50, Avia 16 and Avia 46.

Index

(see also sections for Main Units, Squadrons, Personnel, Airfields)

White, F/O   86
White, W/O   127
Whitehead, F.   14
Williams, S/Ldr   86, 88
Williams, P/O   114
Williamson, F/O P.   112
Wilkinson, Sgt   97
Wilson, F/O   119
Wincott, S/Ldr   104
Wood, W/Cmdr   90
Woodcock, Sgt   92
Woodman, W/Cmdr C. E.   140
Wojcgynski, F/Sgt   84
Wray, G/Capt J.   41, 76, 81

Yaxley, S/Ldr R. G.   108
Ydlibi, P/O   106

Airfields
Abu Sueir   106
Acklington   75, 77, 84, 88
Agartala   121
Amarda Road   121
Ayr   77, 79, 81, 88

Baigachi   123, 128
Banff   101, 102
Berka   110, 114, 115
Bhole Hill   133
Blida   108
Bone   113
Boscombe Down   12, 38, 52, 57

Chiringa   127, 131
Chittagong   124
Chivenor   89, 93, 96
Cholavarum   126
Church Fenton   82
Coleby Grange   42, 77, 79
Coleyweston   78
Coltishall   26, 53, 81, 83
Coomalie   134, 138
Cox's Bazaar   130
Cranfield   88

Dallachy   100–102
Davidson Moor   99

Djidjelli   112
Dohazari   129
Drem   81
Drysdale   134, 138
Dum-Dum   122
Duxford   36
Dyce   91

Edku   102–104, 106, 108, 109, 119
Exeter   79, 87

Farnborough   25, 57
Filton   22, 25, 36, 39, 40, 71
Foggia   118
Ford   83, 85
Forrest Hill   134

Gambut   104, 119
Gaudo   117
Gibraltar   94, 103
Goodenough Island   136, 137
Gosport   34, 38
Grottaglie   117

Hassani   120
Honiley   88

Kanchrapara   121
Kingaroy   142
Kings Cliffe   78
Kiriwina   136, 137
Kirton-on-Lindsey   79
Kumbhirgram   128

Labuan   141, 142
Langham   99, 100
Laverton   44
Leuchars   92, 99
Limavady   91
Luqa   117

Maison Blanche   108, 110, 112
Magrun   110
Manston   100
Martlesham Heath   11, 16
Middle Wallop   74, 77
Meiktila   128, 131

Milligimbi   135, 136
Minneriya   127
Mytkinawa   127
Morotai   141, 142

Nagpur   127
Noemfoor   140
North Coates   92–96, 100
Northolt   26

Phaphamau   123
Port Lyautey   94
Port Moresby   133, 135
Portreath   94
Port Said   106
Predannack   84, 90, 97
Protville   94

Ranchi   128
Ratmalana   125
Rosignano   119

Sanga-Sanga   142
Scorton   81, 84, 88
Silchar   123
St Athan   25, 73
St Eval   98, 102
St Thomas Mount   127

Tain   97
Takali   108
Tanitini   142
Tangmere   25, 73, 74
Tarakan   142
Tebassa   110
Thorney Island   91
Tulihal   128

Valley   86
Vavuyina   127
Vivigani   136

Wattisham   90
Wick   92, 93, 97
Wittering   78, 79

Yelahanka   127
Youks-les-Bains   110